Technology Innovation in Manufacturing

This text identifies and discusses different technology innovation initiatives (TIIs) such as entrepreneurial capability, technology infrastructure capability, organizational culture and climate, and government initiatives. It further evaluates the relationship between various technology innovation initiatives and manufacturing performances using multi-criteria decision-making techniques such as fuzzy set theory (FST), structural equation modeling (SEM), and analytic hierarchy process (AHP). It will serve as an ideal reference text for graduate students and academic researchers in the field of industrial engineering, manufacturing engineering, mechanical engineering, automotive engineering.

This book:

- Discusses technology innovation initiatives such as entrepreneurial capability, technology infrastructure capability, and organizational culture.
- Highlights technology innovation-strategy model in assisting manufacturing industries for enhancing their performance in today's competitive environment.
- Examines the effect of technology innovation initiatives on the performance of manufacturing industries.
- Covers multi-criteria decision-making techniques such as fuzzy set theory, structural equation modeling, and analytic hierarchy process.
- Explores the validation of fuzzy-based technology innovation model through structural equation modeling.

Technology Innovation in Manufacturing

Davinder Singh
Jaimal Singh Khamba
Tarun Nanda

CRC Press
Taylor & Francis Group
Boca Raton London New York

CRC Press is an imprint of the
Taylor & Francis Group, an **informa** business

First edition published 2023
by CRC Press
6000 Broken Sound Parkway NW, Suite 300, Boca Raton, FL 33487-2742

and by CRC Press
4 Park Square, Milton Park, Abingdon, Oxon, OX14 4RN

CRC Press is an imprint of Taylor & Francis Group, LLC

Library of Congress Cataloging-in-Publication Data

Names: Singh, Davinder, author. | Khamba, Jaimal Singh, author. | Nanda, Tarun, author.
Title: Technology innovation in manufacturing / Davinder Singh, Jaimal Singh Khamba, and Tarun Nanda.
Description: First edition. | Boca Raton, FL : CRC Press, 2023. | Includes bibliographical references and index.
Identifiers: LCCN 2022029370 (print) | LCCN 2022029371 (ebook) | ISBN 9781032210278 (hbk) | ISBN 9781032225340 (pbk) | ISBN 9781003272977 (ebk)
Subjects: LCSH: Manufacturing processes--Technological innovations.
Classification: LCC TS183 .S543 2023 (print) | LCC TS183 (ebook) | DDC 670--dc23/eng/20220812
LC record available at https://lccn.loc.gov/2022029370
LC ebook record available at https://lccn.loc.gov/2022029371

ISBN: 9781032210278 (hbk)
ISBN: 9781032225340 (pbk)
ISBN: 9781003272977 (ebk)

DOI: 10.1201/9781003272977

Typeset in Sabon
by Deanta Global Publishing Services, Chennai, India

Dedicated to

my beloved mother

Smt. SURINDER KAUR

Contents

Preface

There is considerable research claiming the importance of innovation in achieving growth and competitiveness for manufacturing industries in the globalized world. Technology adoption in this sector is a growing area of interest in developing countries. Firms which want to gain entry into new markets and/or develop and maintain a competitive edge inevitably require technological innovation. Technological innovation acts as an important factor in competitiveness of a manufacturing unit. It provides a new dimension to industry growth and also has the potential to encourage growth of individual enterprise.

This book highlights the contribution of technology innovation initiatives (TIIs) in manufacturing industries for enhanced manufacturing performance. The empirical analysis of survey reveals that TIIs have yielded considerable significant improvement in manufacturing industries in terms of improved life cycle of products, reduction in cost of production, mean sales profitability, and increase in market shares. These interrelationships between TIIs and manufacturing performance can be used to understand the benefits of these initiatives toward realization of organizational objectives of growth and sustainability.

Author Biographies

Dr. Davinder Singh has worked as an assistant professor in the Department of Mechanical Engineering, Punjabi University, Patiala, Punjab (India) since 2011. He completed his Ph.D. in November 2016 at the same institution. He completed his M.Tech. (Production and Industrial Engineering) at Thapar University, Patiala, Punjab (India) in 2009. He completed B.Tech. in Mechanical Engineering at Giani Zail Singh College of Engineering and Technology, Bathinda, Punjab (India). He has guided more than 30 students for M.Tech. thesis. He has published around 35 research papers in various international journals and conferences. Presently, four students are working under him for their Ph.D. and two for their M.Tech. His main research areas are production and industrial engineering, manufacturing technology, and innovation management. He has held numerous charges in Department of Mechanical Engineering, Punjabi University, Patiala, Punjab (India).

Dr. Jaimal Singh Khamba is working as a professor in the Department of Mechanical Engineering, Punjabi University, Patiala, Punjab (India). He has guided a number of students for their M.Tech. and Ph.D. works. Many students are pursuing their Ph.D. work under him. He has a large number of research projects, conferences, and consultancies to his credit. He has published many research papers in national/international journals and conferences.

Dr. Tarun Nanda is working as an associate professor in the Department of Mechanical Engineering, Thapar University, Patiala, Punjab (India). He has guided a number of students for their M.Tech. and Ph.D. works. Many students are pursuing their Ph.D. work under him. He has published many research papers in national/international journals and conferences.

Acknowledgments

I owe a deep sense of gratitude and indebtedness to my family members and friends for their inspiration, blessings, and endeavor to keep my morale high.

I put on record my sincere thanks to the manufacturing organizations for their help and support in conducting case studies.

In addition, I thank the editors and publisher of this book.

Chapter 1

Technology innovation and its significance

1.1 INTRODUCTION

This research book focuses on the function of technology innovation (TI) in the performance enhancement of manufacturing industries. The manufacturing sector plays an important role in most developed as well as developing economies because of its number, variety, and involvement in all the segments of economy (Todd and Javalgi, 2007). But this sector, especially in developing countries, is exposed to intense competition because of the accelerated process of globalization (and various provisions of the WTO regime), which brings out the necessity for the units in this sector to develop competitiveness for their survival and growth. These firms are generally constrained in terms of resources such as technology, finance, marketing, and human resources in developing countries. The capability of the manufacturing sector of any country to compete in the global marketplace depends on its access to these kinds of resources. Units that have superior access to these resources are able to exhibit improved economic and innovative performance (Sikka, 1999; Bala Subrahmanya, 2007).

The majority of economic structures are composed of manufacturing enterprises, and the majority of employment is concentrated in this sector (Zeng et al., 2010). It always remains in the forefront of the economic policy debate, and governments consider these firms as the engines of development and growth, especially in developing countries (Peres and Stumpo, 2000). While the definition varies, there is a need to look closely at the characteristics and behavior of this sector, in view of their perceived significance in job creation and economic growth (Khayyat and Lee, 2015). It is little known, for example, regarding the volume of firms involved in innovative activity and about the nature of that activity (Mothe and Thi, 2010).

Manufacturing firms, in general, must take innovative initiatives for their growth and survival in the era of globalization (Edoho, 2016). These firms rely on their external linkages and/or on their internal technological capabilities as the sources of innovation. With the current pace of technological development, access to resources and knowledge from outside the organization is becoming more and more important (Edwards and Delbridge, 2001).

DOI: 10.1201/9781003272977-1

1

1.2 INNOVATION TYPOLOGY

The traditional categorization of innovation as radical or incremental is seen as incomplete or too simplistic (Henderson and Clark, 1990; Garcia and Calantone, 2002), and there is a need to take into account the continuum between these two extremes. Garcia and Calantone (2002) developed a tridimensional model mapping innovation according to (i) the level of innovation, (ii) technological versus marketing innovation, and (iii) newness to the enterprise versus newness to the industry. The model generates three categories of innovation: radical, significative, and incremental. Tidd and Hull (2006) offer a classification based on the impact of change and whether it involves a specific component, a sub-system, or the whole system (the enterprise, the sector, the industry). The literature has produced different factors to study the newness and the perspectives from where innovation is considered. Garcia and Calantone (2002) identified, through a systematic literature review, at least six perspectives of newness in a set of empirical research from 1979 to 2000: newness to the world, to the enterprise, to the industry, to the market, to the customer, and to the scientific community. Another model from Henderson and Clark (1990) distinguishes innovation at the product component level and innovation at the architecture level. This model introduces two dimensions: (i) the innovation's impact on products components and (ii) the impact on the linkages between components. This model enhances the dichotomy incremental/radical by recognizing intermediate level innovations such as modular innovation and architectural innovation.

We can see from these few examples that most of the innovation typology models include a dimension about the degree of innovation, but they are very different when it comes to the other dimensions involved. The factors used to identify innovation types don't have the same reach and depth in the research, leading to fragmentation and difficulty to make use of the results (Table 1.1).

1.3 TECHNOLOGY INNOVATION DEFINED

Innovation is supposed to be the means to create new markets and growth opportunities (Lagrosen, 2005). Technological innovation is considered one of the major factors for an organization's continuing success and competitive advantage (Corso et al., 2001; Du Plessis, 2007). Numerous studies have widely recognized the significance of technological innovations (Dewar and Dutton, 1986; Gloet and Terziovski, 2004; Lundvall and Nielsen, 2007). It can generally be defined as the implementation of an initiative or a conduct which is entirely new for a firm (Damanpour and Gopalakrishnan, 2001). Innovation is an elusive concept in which various key issues require consideration. The seminal study on economic development and entrepreneurship by Schumpeter (1939) defined innovation as the establishment of a new production facility.

Table 1.1 Innovation Typologies Review

Authors	Innovation Types	Dimensions
Schumpeter (1942)	Continuous/discontinuous	Level of change
Schumpeter (1942)	Product/process production methods/ market/industrial organization	Innovation categories
Christensen and Overdorf (2000)	Continuous/disruptive	Level of change
Bessant et al. (2005)	Continuous/discontinuous	Level of change
Tidd and Hull (2006)	Incremental/radical	Level of change
Gardiner and Rothwell (1985)	Platforms/derivatives	Level of reuse
Garcia and Calantone (2002)	Incremental/radical/significative 'really new'technology/marketingmacro (market, industry)/micro (enterprise)	Level of changeInnovation categoriesLocus of impact
Garcia and Calantone (2002)	Newness: to the world/industry/ market/customer/enterprise/scientific community	Locus of impact
Francis and Bessant (2005)	Product/process/position/paradigm (business model)	Innovation categories
Choffray and Doray (1983)	Repositioned product/reformulated product/original product	Innovation categories Level of Newness perception
Henderson and Clark (1990)	Incremental/modular/architectural/radical	Impact on products componentsImpact on the linkages between components
Edwards and Delbridge, (2001)	Technology product and process/ service/marketing/organizational	Innovation categories
Fernez-Walch and Romon (2006)	Local/global	Locus of impact
Tidd and Hull (2006)	Component/sub-system/system (enterprise, sector, industry ...)	Locus of impact

Technology innovations can take many forms such as product innovations, process innovations, administrative innovations, technical innovations, radical innovations, and incremental innovation. On the other hand, Mavondo et al. (2005) categorized innovations into three types, such as product, process, and administrative innovations. However, two discrete forms of innovations, namely process and product innovations, have been recognized and widely accepted by other researchers (Chuang, 2005; Tidd, 2001; Prajago and Sohal, 2004).

Product innovation includes the formations of new products and services that aim at meeting the demands and expectations (Damanpour and Gopalakrishnan, 2001), whereas process innovation includes implementation of new and modified production and/or delivery techniques, by means of the amendments in methods, tools, and/or software (Bi et al., 2006; Cassimen et al.,2010). Therefore, product and process innovations have been believed as the two essential elements that made up technological innovations (Chuang, 2005; Cooper, 1998; Damanpour and Gopalakrishnan, 2001).

Comparing the various types of innovations, it has been observed by many researchers that technological innovation is the most beneficial form of innovation for the manufacturing firms because of its ability to increase performance, provide added value, resolve problems, and also help to develop a competitive advantage for the firm (Cooper, 1998; Tidd and Besant, 2009; Zaugg and Thom, 2003). These firms are dependent on technological innovation for producing and manufacturing quality products (Bi et al., 2006).

Some of the important definitions of innovations/technology innovations provided by various researchers have been summarized in the following paragraphs:

Harris and Robinson (2002) refer to innovation as a 'broad definition of innovation ... including not only the "hard" technologically determined definition ... but also the organizational and managerial practices'.

Zairi (1994) in discussing a definition of innovation states: 'what makes innovation challenging is the fact that it is very difficult to agree on a common definition, and also to decide which firms are the most innovative and how to quantify innovation activity?'

'Innovative companies are especially adroit at continually responding to change of any sort in their environments and are characterized by creative people developing new products and services' (Peters and Waterman, 1982).

Tushman and Nadler (1986) refer to product, process, and technological innovations. They state that 'innovation is the creation of any product, service or process, which is new to a business ... the vast majority of successful innovations are based on the cumulative effect of incremental change in ideas or methods'.

Thus, innovation in organizations covers both organizational and technological perspectives of innovation (Mosey et al., 2002). Therefore, these perspectives must be combined in theoretical as well as practical approaches to enhance innovation in manufacturing industries.

1.4 IMPORTANT SOURCES OF TECHNOLOGY INNOVATION

In developing countries, it is a matter of significance that manufacturing enterprises undertake technological innovations because not much light has been thrown on this aspect so far. Manufacturing industries in the developing world, in general, are considered to be having a low level of

skills and technology as compared with developed ones. Still these industries may have managed to develop in-house technological capability by
employing exclusive personnel and equipment, and therefore carry out
innovations on their own, whereas others may not have sufficient technological capability in-house and therefore are dependent on external support and resources.

Soderquist et al. (1997) proposed the use of cross-functional teams,
investment in R&D, developing closer working relationships with preferred customers, and the propinquity of customers as important sources
of innovation and suggest that the sources of innovation ideas can be both
internal and external. The internal sources consist of internal R&D, marketing group, top management, and manufacturing, whereas the external
sources include clients or customers, competitors, collaboration with other
companies, suppliers, machinery and equipment, university or research
institutions, acquisition of new equipment, consultants, internets, and professional journals (Bommer and Jalajas, 2004).

Sources of innovation ideas can come from research and from transnational corporations through subcontracting. Findings of the study
conducted by Bala Subrahmanya (2007) indicated that support received
through subcontracting is beneficial as it enhanced technological innovations of manufacturing industries in India and facilitated their economic
performance. Xie et al. (2010) reported 'customer' as the most important
cooperation partner for innovation in manufacturing industries.

As shown in Figure 1.1, manufacturing industries in developing industries are usually dependent upon traditional sources for technological innovations. These sources include their professional and personal networks
from inside and outside of the business. It comprises intermediaries and
consumers, traders or crafts-persons, or wholesalers who provide inputs
on various aspects such as shapes, design, and characteristics of products.
Importing new technology from abroad is restricted very much in case of

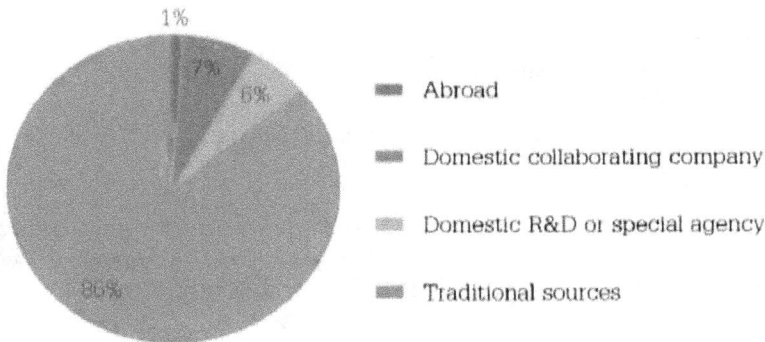

1%

7%
6%

86%

■ Abroad

■ Domestic collaborating company

■ Domestic R&D or special agency

■ Traditional sources

Figure 1.1 Sources of technology innovation for manufacturing industries (Fagerberg and
Verspagen, 2009).

manufacturing in developing countries. Few firms collaborate with domestic firms and research institutes for technological innovations (Fagerberg and Verspagen, 2009).

1.5 SIGNIFICANCE OF TECHNOLOGICAL INNOVATION

Manufacturing industries in developing countries like India are presently facing enormous problems that are required to be addressed immediately. These problems are related to technology, scale of operation, logistics, production techniques, finance, local and global competition, marketing, and meeting customer expectations. The answer to these problems can be strategically handled by our manufacturing industries by adopting technological innovations in all activities related to their day-to-day operations.

Interest in the contribution of technology innovation to national economies has been increasing (Romer, 1994; Grossman and Helpman, 1994; Barro and Sala-I-Martin, 1995; Shefer and Frenkel, 2005). Innovation can lead to enhanced market share, greater production efficiency, increased revenue, and higher productivity growth (Shefer and Frenkel, 2005). It enables firms to offer a wide variety of differentiated products that can improve financial performance (Zahra et al., 2000). Keizer et al. (2002) suggest that technology innovation contributes to economic growth and is one of the most important means through which manufacturing industries can remain competitive. It also enables firms to achieve better financial performance by providing a large variety of valuable, rare, inimitable, and differentiated products (Zahra et al., 2000).

Following are the rationales for adopting TI:

1. TI has the ability to increase performance, solve problems, add value, and develop competitive advantage (Cooper, 1998); therefore, it is the most significant among the various types of innovations.
2. Manufacturing firms rely on TI to produce new high-end products (Bi et al., 2006).
3. TI is associated with modifications in the present products and processes (Robert, 2007; Bi et al., 2006).

Technology innovation is considered to be necessary for the survival as well as growth of individual firms. The pressure of competition, which at one time makes some firms innovative, pushes others, who are driven by the character of survival and growth, to seek to catch up and remove the original competitive advantage of innovating firms (Hayland, 2004). Technological innovations in the area of both product and process development are taking place at a very quick pace. It is the only way through which a country can survive (Stoian et al., 2016) as it is not only the means to progress quality, increase production, and develop new products, but also to increase competitiveness, expand exports, and ultimately ensure continuous economic growth (Chaston, 2012).

Cainelli et al. (2004) and Regev (1998) found that innovating firms had greater labor productivity and sales growth as compared with non-innovating firms. A study on British manufacturing industries conducted by the Cambridge Small Business Research Centre (1999) highlighted that 80% of the companies that developed innovation activities improved profits, new markets penetration, and market share. Hughes (2001) found that highly innovative British firms enhanced their profit margin. Hsueh and Tu (2004) showed that innovation positively affected earnings of Taiwanese firms. Bhaskaran (2006) found that manufacturing industries that focused on sales and marketing innovations were able to compete with large companies successfully in Australia.

Heunks (1998) observed that benefits derived from technology innovations may not be visible in the short term, but may take time to be realized. Olav and Leppalahti (1997) indicated that innovating Norwegian firms experienced higher profit margins than non-innovating firms.

Major achievements of innovations in manufacturing industries in India were observed in terms of enhanced competitiveness, improved quality, reduced rejection, improved product designs, increased output, growth in the local and domestic market and penetration in the international market, growth in size over time, etc. (BalaSubrahmanya, 2007).

Various studies have suggested that technology innovation is one of the important factors that lead to improved business performances in terms of indicators such as increased market share, improved product quality, improved health, reduced materials cost, and safety and environment. Lehtimaki (1991) found that innovated new products contributed toward enhanced total sales and profits. Roper (1997) observed that the output of innovative firms grew significantly faster than that of non-innovators as innovated products contributed to the faster growth of the former. Engel et al. (2004) ascertained that sales turnover of innovative firms grew faster as compared with non-innovative firms. Effects of innovation were felt in terms of both product-oriented results such as improvement in quality of goods, increased range, and services on goods and services, and process-oriented results such as improved production capacity and increased production flexibility (Lumiste et al., 2004).

Several studies found a strong positive relationship between innovation and growth (Roper et al., 1996; Roper, 1997; Moore, 1995). Geroski (1994) suggested two dimensions of the relationship between innovation and growth. First, the production of new products or processes strengthens a firm's competitive position, and second, the process of innovation improves the firm's internal capabilities, making it adaptable and flexible to varied market pressures.

1.6 TECHNOLOGICAL INNOVATION IN MANUFACTURING INDUSTRIES

Innovation was traditionally viewed as taking place generally within a single firm. But with the increase in mobility and availability of knowledge workers, flourishing of the internet, invention of venture capital markets, and the increased range of possible external suppliers in the present environment have diluted the efficiency of traditional innovation systems (Chesbrough, 2003). Companies now desire to incorporate in their business models the commercialization of not only their personal ideas but also the external ideas. Therefore, the concept of open innovation has consequently been extended to different perspectives (Christensen et al., 2005).

Innovation is one of the most broadly discussed concepts in the literature of economics. Economists have studied both the micro and macroeconomic aspects and determinants of innovation, while managerial studies generally tend to focus on specific variables of innovative organizations (Fagerberg and Verspagen, 2009). A linear model of innovation is provided by most of the business and economic studies. These models demonstrate that basic research transmitted to industries and from there to the market initiates in research institutes, universities, and research laboratories (Godin, 2006).

There is a considerable research claiming the importance of innovation in achieving growth and competitiveness for manufacturing firms in the globalized world. Technology adoption in the manufacturing sector is a growing area of interest in developing countries. There are established theories in technology adoption that have been extensively applied to western context. But, these theories have not been generally applied to developing country context. Technology adoption is also crucial for the growth of business in the private sector (Edwards and Delbridge, 2001). Firms which want to gain entry in to new markets and/or develop and maintain a competitive edge inevitably require technological innovation (James et al., 2014). To improve quality, improve product shapes/dimensions, reduce costs, increase the range of products, and respond to market challenges are the main reasons for innovations in manufacturing firms (Subrahmanya, 2005).

1.7 CONCLUDING REMARKS

This chapter has given an overview of TI and its significance in maintaining the competitive position of manufacturing industries especially in developing countries. For the growth of an economy, contribution of the manufacturing sector is immense, and at the same time, this sector is facing extreme constraints and pressures to maintain its competitiveness in developing countries in the era of globalization. Competition from multi-national companies, lack of finance, recession, and low demand are becoming exposed problems to this sector. In the present competitive environment, this sector is required to plug into the new market opportunities and to confront the increasing competition from developed economies (Thampy, 2010).

Enterprises are required to bring modifications in the way of doing routine activities or innovation in technology to the business environmental policy goals. These changes and modifications require technology innovation as one of the main processes in organizational objectives. The next chapter reviews the existing literature with regard to the technology innovation initiatives required by the manufacturing sector in order to enhance their manufacturing performance.

Chapter 2

Globalization and its impact on technology innovation

2.1 INTRODUCTION

Technological developments, both inside and outside of factories, have impacted the manufacturing industry's globalization – the processes by which businesses and other organizations develop international influence or start operating internationally. Ever since the first industrial revolution, industrialization has impacted international business. In particular, advances in transportation and telecommunications have had a huge impact. With increasing trade and communication, more and more companies are extending their reach across land and sea (Greco et al., 2015).

In fact, the modern manufacturing supply chain is centered around globalization. Every day, goods are moved across the globe on shipping lines, on freight forwarders, and by air. Business activities, including outsourcing of logistics, facilities management, professional services, and maintenance, can all be international processes (Kristiansen, 2003).

2.2 GLOBALIZATION AND FOURTH INDUSTRIAL REVOLUTION

The technological innovation brought about by the Fourth Industrial Revolution has spurred on the transformation of production, mainly in two aspects. The first one is to promote the upgrading and integration of already established industries. The wide application of new technologies encourages the intelligent development of industries and accelerates the structural upgrading and restructuring of global industrial chains. With the penetration and integration of new technologies in different industries, new industrial models are taking shape. Second, new industries are emerging. For example, the continuous development of artificial intelligence technology has created a new industry revolving around robotics. Research developments in lithium batteries and charging piles have catalyzed the rapid growth of the new energy industry (Shang et al., 2010).

DOI: 10.1201/9781003272977-2

However, the Fourth Industrial Revolution is facing new challenges. At present, the world is witnessing rising protectionism and unilateralism, intensifying trade and investment disputes, a severely handicapped multilateral trading system, and mounting systematic risks and other global challenges. In 2019, many international institutions, such as the International Monetary Fund and the World Bank, lowered their global economic growth forecasts. The risks and uncertainties hovering around global economic development have increased significantly (Khayyat and Lee, 2015).

With every major industrial and technological change, the characteristics of globalization have been altered. In 2011, the term Industry 4.0 was introduced by the German government and Siemens. Industry 4.0 shifts manufacturing away from analog and mechanical technologies and toward all things digital.

As information technology and operational technology converge, companies are beginning to find new ways to connect. Data collected from suppliers, customers, and the enterprise can be aligned with detailed production information, which means processes can be fine-tuned in real-time (Pasadilla, 2010). The digital and physical worlds have become irrevocably linked, with machines, systems, and people able to exchange information and automatically adjust. Industry 4.0 is not only revolutionizing manufacturing processes but also having a powerful impact on the model of globalization, by changing the workforce and increasing the ease of access to services (Tidd, 2001).

2.3 POSITIVE AND NEGATIVE CONSEQUENCES OF GLOBALIZATION

While globalization acts as a medium for human progress, it is also a disordered process which offers both disadvantages and benefits to people across the world.

Positive consequences of globalization include:

1. Developments in local productivity promoting prosperity.
2. The movement and contribution of knowledge, information, and expertise.
3. The development of international standards for health and education.
4. Enhancing the range of goods accessible to the world market and offering a bigger scope of markets for the internationally sourced products.

Negative consequences of globalization include:

1) Slaughter of employment in developed countries.
2) A flow toward a more homogenized society and culture internationally.

3) Local economies are more exposed to fast alterations in the international economy.
4) Centralization of power in the hands of the large transnational corporations.
5) Geographical location of the industry in less developed economies often leads to environmental degradation.

Globalization has a negative effect on the growth of small firms measured in terms of production, employment, exports, and number of units. A decrease in the growth of employment generation and number of units in the post liberalization period is an issue of serious concern for planners and policy-makers (Sonia and Kansal, 2009).

2.4 EFFECT OF GLOBALIZATION ON TECHNOLOGY INNOVATION IN MANUFACTURING INDUSTRIES

Globalization is the process of combining different economies of the world without generating any obstructions in the free flow of technology, capital, goods and services, human capital, or even labor. Therefore, it indicates internationalization plus liberalization, by which the world has turned into a small global village (Lahiri, 2012).

Today, globalization involves various features, but the following three are the main factors for driving global economic integration:

1. Internationalization of production along with changes in the structure of production.
2. Widening and deepening of international capital flows.
3. Expansion of international business in trade and services.

Globalization in India is normally considered as 'integrating' the economy of the nation with world economy. The actual force to the globalization process was offered by the new economic policy established by the Government of India in July 1991 after the request of IMF and World Bank. It has led to an 'Unequal Competition' – that is competition between 'MNC's and Indian enterprises'. Small scale sector is considered as a vital constituent of overall industrial sector of Indian economy. It forms a central part of Indian industry and contributing to an important proportion of employment, production, and exports. Thus, it is required to study and analyze the effect of globalization on small scale sector in India (Chandraiah and Vani, 2013).

Globalization has brought in a new pattern in the economic environment of Micro, Small, and Medium Enterprises (MSMEs). The rising integration between countries no doubt resulted in several opportunities in terms of wide customer base, technology transfer, new markets, managerial skills enhancement, etc. for this sector but posed new challenges of reduced product life cycle and growing international competition. This transformation in the global economy as a whole has enforced these firms to strengthen their manpower, technology, and research base so as to survive in the global market by producing quality and cost-effective products. Therefore, strategic attitude has become a matter of necessity and not a choice (Taneja, 2013; Subrahmanya, 2015).

2.5 NEED FOR TECHNOLOGY INNOVATION IN THE ERA OF GLOBALIZATION

Technological innovation acts as an important factor in competitiveness of a firm. At the macro level it provides a new dimension to industry growth and also has the potential to encourage growth of individual enterprises at micro level (Liyanage, 2003). MSMEs are usually considered as more flexible, found to develop and/or implement new ideas, and adapt themselves better. Simple organizational structure of MSMEs, low risk, and receptivity and flexibility are some of the important features which facilitate them to be innovative. Therefore, MSMEs have the unrealized innovation potential across other industries (Becheikh et al., 2006; Chaminade and Vang, 2006).

The present work also provides a complete technology innovation implementation program for the MSME sector to improve its performance in this competitive world. In the present study, both primary and secondary kinds of data have been collected and used so that suitable analysis is carried out to justify the problem statement.

2.6 CONCLUDING REMARKS

Globalization has brought in a new pattern in the economic environment of manufacturing enterprises. The rising integration between countries has no doubt resulted in several opportunities in terms of wide customer base, technology transfer, new markets, managerial skills enhancement, etc. for this sector but has posed new challenges of reduced product life cycle and growing international competition. This transformation in the global economy as a whole has enforced these firms to strengthen their manpower, technology, and research base so as to survive in the global market by producing quality and cost-effective products. Therefore, strategic attitude has become a matter of necessity and not a choice.

Chapter 3

Technology innovation initiatives in manufacturing industries

3.1 INTRODUCTION

Micro, Small, and Medium Enterprises (MSMEs) are universally recognized as a powerhouse of growth and backbone of any economy and also a major contributor toward GDP of any country and employment generation. Their importance in growth of economy, upliftment of the poor, resource generation, and innovation in technology is known globally (Subrahmanya, 2015). With the offset of industrial revolution and globalization, the MSME sector has grown phenomenally but it is also facing ruthless competition from its counterparts both globally and locally. Surviving such a stiff competition requires MSMEs to continually strive for newer technologies and think innovatively. These organizations need to ascertain ways to stay afloat and competitive in this ruthless competition. One strategy can be to work innovatively so as to bring breakthrough products and creative ideas to the market. Organizations need to identify ways in which they can stay innovative and outperform their competitors (Fugate and Kinicki, 2008).

Innovation in organizations is a prerequisite for economic and technological growth of any organization. Identifying technology innovation initiatives can help managers to proliferate the process of economic growth.

3.2 TECHNOLOGY INNOVATION INITIATIVES

A study about innovation in MSMEs has investigated the scope of innovation, like the connection among innovation and firm routine, social environment and innovation of the small firms (Subrahmanya, 2015), and small firms and societal network (Freel, 2005; Sharif et al., 2012; Subrahmanya, 2015). It is observed that usually MSMEs don't innovate in a formal manner, and 'learning by doing' is the most ordinary style of innovation. Therefore, a large number of scholars considered innovation in MSMEs as a casual process of innovation (Kristiansen, 2003). Innovation in MSMEs has also been considered the same, as a characteristic for the entrepreneur, deriving from the motivation and vision of the entrepreneur (Abereijo et al.

DOI: 10.1201/9781003272977-3

2009). From enormous studies conducted by various researchers, it is found that *Entrepreneurial Capability, Technology Infrastructure Capability, Organizational Culture and Climate*, and *Government Initiatives* are the main input factors improving the technology innovation in small firms.

3.2.1 Entrepreneurial capability

French economist Richard Cantillon, in the early 18th century, put forth the term entrepreneur. He formally defined it, in his writings, 'as the agent who buys means of production at certain prices in order to unite them into a new product'. Further, he defined the term entrepreneurship as 'self-employment of any kind where the entrepreneur is the bearer of uncertainty and risk'.

Soon after that, the French economist Jean Baptiste Say defined entrepreneur as 'somebody who shifts economic resources out of an area of lower to an area of greater yield and higher productivity'. He added Cantillon's definition together with the idea that 'an entrepreneur is one who builds a single productive organization by bringing other people together'. But according to Peter Drucker, Say's definition is not able to tell us exactly who the entrepreneur is. And since Say coined the term roughly 200 years ago, there has been a deficiency of consensus over the definition of entrepreneurship and entrepreneur (Kim, 1988).

A critical role of 'innovation' to the entrepreneur was assigned by Joseph Alois Schumpeter, for the first time, in 1934, in his 'magnum opuses', which is a theory of 'economic development'. Schumpeter considered 'economic development' as a distinct dynamic change. Such discontinuous dynamic changes are done by entrepreneurs by applying new combinations of the factors of production, i.e., 'innovation' (Schumpeter, 1934).

Entrepreneurship, as demonstrated by the characteristics of the entrepreneur, is considered to be essential to the characteristics of MSME performance by some researchers. Such a claim was mainly popular among studies of the entrepreneurial firm, in which the entrepreneur acted as a founding and leading role model in the progress of the enterprise. Many researchers have tried to explore the various characteristics of entrepreneurs affecting the performance of MSMEs, including the entrepreneur's background as well as demographic characteristics like gender, education, age, and ethnic origin. The direct and indirect assistance of the opportunity, entrepreneur's relationship, and human competencies influence the long-term performance of a small firm by means of organizational capabilities and competitive scope (Man et al., 2012).

Mainly, entrepreneurship is a grouping of three dimensions: risk-taking, innovativeness, and pro-activeness. Entrepreneurial orientation refers to 'a firm's strategic orientation, confining specific entrepreneurial features of decision-making styles, practices, and methods'. Innovativeness reflects an affinity to support new ideas, novelty, experimentation, and creative processes, thus departing from established technologies and practices. Entrepreneurship perhaps has constructive performance implications for any firm. The decrease in product and business model lifecycles makes future profit streams from existing operations doubtful and at the same time businesses require to continually look for new opportunities. Innovative companies that create and introduce new products and technologies can produce a surprising economic performance (Wiklund and Shepherd, 2003).

3.2.1.1 Education level of entrepreneur

Today's concept of entrepreneurship education has a wide meaning, which includes various factors like economic, social, and cultural. Therefore, entrepreneurship education is a social and dynamic process in which individuals, either alone or in collaboration, recognize opportunities for innovation and work upon these by converting ideas into targeted and practical activities, whether in economic, social, or cultural context (Greco et al., 2015; Kristiansen, 2003; Shang et al., 2010).

Education is essential in sharpening the tacit capabilities of an entrepreneur. Although it has been argued that the formal education system is not sufficient for learning entrepreneurship, the expansion of entrepreneur's awareness of the business environment around him makes education essential. Education becomes a helpful tool to the entrepreneur in circumstances where the problem faced requires decision making and taking action outside the sphere of normal business operations. For making quick and right decisions, having a wide knowledge of things is essential, if not inevitable for a small scale entrepreneur (Alkali, 2012).

3.2.1.2 Entrepreneur training

Due to their small size and resource limitations, MSMEs particularly face numerous problems. Availability of finance is a major problem for small firms in starting a new project. Most of them also face problems with regard to lack of finance for expanding an established business. Due to their inadequate resources, they suffer more from administrative burdens and red tapism as compared to larger enterprises. They often lack in developments regarding communication and information technologies, and come across difficulties in finding competent staff as well as providing them with appropriate training and education. Moreover, it can also be problematic to find successors for retiring business owners (Khayyat and Lee, 2015).

The levels of technology innovation and adaptation that can be accomplished by a firm are largely governed by the quality of the workforce in the labor market and the training availed by the employees. Training, both on-the-job and final, is one basis of technology learning that supports and substitutes firm level technology competencies (Pasadilla, 2010). Training is aimed at exposing the workers to new technologies and/or increases their performance, enhancing technological capabilities, production, and efficiency over time (Kor and Mesko, 2013).

Entrepreneurship training is one of the most complicated problems hindering small firm growth. In spite of the increasing contribution of non-governmental organizations and other small enterprise training agencies in the informal sector, their training programs have had little or no impact on the change of attitude and attainment of entrepreneurial skills to target beneficiaries. Reasons for this fact have not been well discovered and neither do we have empirical evidence to explain the same (Tidd, 2001).

3.2.1.3 Technical competencies of entrepreneur

The business operation is constantly changing with fast technological advancements which are considered to be very complex in a competitive business environment. An entrepreneur is expected to act together with these environmental factors which require him to be highly proficient in various dimensions like attitudinal, intellectual, behavioral, managerial, and technical aspects (Spokane, 1991). Therefore entrepreneurs are permanently challenged to organize a set of competencies to be successful in their entrepreneurial endeavors, growth, and/or survival. Some of these competencies are inherent, while others are attained in the process of learning, training, and development (Jennings and Cox, 1995). Troilo (2014) also defined competencies as knowledge, skills, and personal characteristics of an individual.

Entrepreneurial competencies are defined as fundamental characteristics possessed by an individual, which result in new venture creation. Since maintenance of machinery is considered as one of the major elements that comprises technology innovation and adaptation capability (Lall, 1992), it can be thought that the firms in the sample had reasonably high technological development competencies. In small scale firms, some of their technology innovations originate often during the carrying out of preventive and routine repairs and maintenance. Majority of the innovations that originate by this method are alterations to the machinery that are undertaken either to avoid break downs or improve on the machine performance. In-house maintenance and repair potentials are therefore critical to firms because they offer the foundation to shop floor innovations, which are considered as the most important types of technological advancement that take place in MSMEs (Tidd, 2001).

Within an enterprise, there are numerous factors that decide its capacity to carry out in-house repair and maintenance. The most critical one, however, is the levels of knowledge, skills, and experience of the entrepreneur (Hall, 2002).

3.2.1.4 Work experience of entrepreneur

Entrepreneur's ability to innovate also depends upon the previous work experience which is another important attribute. Work experience is useful to the entrepreneur as it helps him in the accumulation of technical know-how as well as other skills required for innovation (Lall, 1992). The longer the period for which an entrepreneur has worked, the more skills and experience he ought to have acquired from the job which widens his entrepreneurship intelligence (Fugate and Kinicki, 2008).

Establishment of new firms is one of the channels which affect economic development of any country. Using firm-level data for eastern and western Europe, it is discovered that entry regulations obstruct the creation of new firms, while regulations that promote access to finance boost entry (Barel, 2013). The study also suggests that in several cases a poor business environment might affect the performance of MSME sector, because limitations and market deficiencies reduce competition and slow down firm growth.

A comparison of the UK and Italy illustrates this effect. In Italy and the UK where entry costs are 20% and 1.4% of GNP respectively, there are a lot of small firms with slower growth. The problem in Italy is that the MSME sector has numerous old and incompetent firms compared to its UK counterpart. Certainly, firms start at a larger scale in Italy, but grow more slowly as compared to the firms in the UK which are about twice as large by age 10 (Beck and Kunt, 2006).

Customer preferences and business environment will always have a change to be more complex as well as dynamic in nature. Firms can only survive by making internal changes to balance the changes that occur in market. To deal with the changing customer preferences and changing business environment, the idea of market orientation has been introduced by many companies, which is one of the significant developments in marketing studies (Egbetokun et al., 2012).

3.2.1.5 Financial schemes and loan procedure

The Indian MSME sector has a great impact on their growth and development because of a lack of adequate access to finance. MSMEs are not capable of accessing the capital markets for their financial requirements, and therefore, banks act as an important source of funding for these firms. In recent years, policy makers and governments have been giving significant attention to facilitating the development in the MSME sector, because this sector gives a good foundation for entrepreneurship as well as development in the economy (Thampy, 2010). Limited access to formal sources of external finance leads to reduced contribution of MSMEs toward economic growth as small firms face larger growth constraints compared with large firms. Institutional and financial development helps improve MSMEs' growth constraints and enhance their access to external finance and hence levels the playing field among firms of different sizes (Egbetokun et al., 2012).

Financing obstructions are preventing small firms from reaching the optimal size. It is also highlighted in literature that small firms devote a smaller share of their working capital and investment with formal financial sources as compared to large firms. There is the effect of financial market structure to lessen MSMEs' access to financing tools and techniques to overcome small firm's financing constraints. The introduction of transaction-based

MSME financing tools like factoring and credit scoring, on the other hand, has emphasized the advantages of large banks that provide finance to small recognized firms (Beck and Kunt, 2006; Okrnglicka, 2014).

There is a lack of simplicity concerning the financial conditions of MSMEs in the Indian financial system (Berry, 1998). Unless fair and thorough information on small firms is available, banks would hesitate to take the risk and may desire to lend to comparatively larger firms to comply with regulation, consequently leaving smaller firms constrained for capital. Improving the quality of financial information of small firms is an essential requirement for increasing the flow of funds to the MSME sector, as the decisions on loan finance are also influenced by quality of information (Das, 2008).

3.2.2 Technology infrastructure capability

Broadly, technology infrastructure is defined as elements of an organization's technology base that is being originated outside the traditional boundaries of the firm and which is then used by a majority of the firms in the sector (Wang et al., 2007). A number of studies illustrate that innovation barriers are related to organizational culture, institutional constraints, cost, human resources, flow of information, and government policy (Hall, 2002; Acharya, 2008; Troilo, 2014).

Competitiveness of any industry depends upon the technological capability to enhance its economic performance. Technological capability has been termed as an accumulation of technological knowledge gathered by an organization over time. It reflects the capability of not only responding speedily through modifications in products and processes, but also providing the cutting edge in competing with other firms through innovation activities. The gathering of this technological knowledge is obtained by personal mastery of organizational learning and new knowledge (Masurel et al., 2010).

The idea of technological capability has been more broadly used in the present competitive environment. Technological capability occurs cumulatively and gradually in a firm. Generally, it begins with simple routine activities, through more complex duplicative and adaptive activities requiring searching operations, leading to the most innovative activities which are based on the more formalized research (Lall, 1992). Innovation capability is one of the important characteristics of technology capability (Romijn, 2001). Drejer (2003) has proved the association between organizational learning and innovation. Numerous studies have proposed new theories for knowledge management and product development.

3.2.2.1 Material resources

It is necessary that quality products should be produced at reasonable prices as it determines the price of a product. Therefore, the main focus should be to motivate the small scale sector to manufacture high quality goods/

products by providing them with necessary financial assistance, technical guidance, raw materials, marketing assistance, etc. In the process of technology innovation, raw material plays a significant role and therefore raw material should be available at reasonable price to enhance innovation performance of an organization (Spokane, 1991; Koch and McGrath, 1996).

Small scale industries are weak in financial position and are faced with the problems of shortage of raw materials like iron and steel, grade coke, pig iron, chemicals etc. These industries have to use the services of intermediaries, but such an activity results in elevated costs and is disadvantageous when raw materials are imported, because the profit margins of intermediaries are rather high (Kim, 1988).

Types of raw materials used by industries depend upon the requirement of any industry. There are some industries that utilize indigenous raw materials while others are based on imported raw materials (Beer et al., 1984). With small scale industries, the non-availability of raw material in adequate quantities remains the main problem. The lack of scarce raw materials is expected to continue in future also. Therefore, the policy for the development of MSMEs should be to encourage this sector based on local and indigenous raw materials whereas those based on imported raw materials should be discouraged. However, it should be the accountability of the state to fulfill all the requirements of the existing industries by guaranteeing liberal allocation of raw materials from the issuing import licenses and state depots for reasonable amount required for production (Subrahmanya, 2005).

3.2.2.2 Research and development expenditure

R&D is categorized into three types: basic research, i.e., experimental work carried out without any specific use or application in view; applied research which is an organized examination that is carried out with a commercial or practical objective; and development research which is usually a work meant at perfecting an invention i.e., translating basic and applied research results into new and/or improved processes and products.

Therefore, it is also known as 'any creative and systematic activity undertaken to enhance the stock of innovations and inventions' (Bwisa and Gacuhi, 1997). In majority of the cases, R&D is undertaken in well-recognized laboratories whether in universities, R&D institutions, or even in industry. Being a 'systematic' activity, it is generally well organized with its own allocated resource amenities, like finance and manpower. However, in developing countries this ideal condition does not always exist especially for MSMEs. While the type of R&D management and organizational structure can be termed to be 'formal' and is both innovation- and invention-oriented in the former case, it is more 'informal' and basically innovation-oriented in the latter case of MSMEs in developing countries (Drejer, 2003).

With the inadequate financial, capital, and human resources at the disposal of MSMEs, it becomes very important for these firms to collaborate with other institutions in undertaking their R&D activities. Linkage which is defined as 'a network of relationships of productive units, surrounded by a framework of inter-industry and intra-industry externalities' is an idea that is exceptionally beneficial in R&D activities in developed economies for manufacturing firms. However, in developing countries, MSMEs' collaboration and R&D linkage are still very low (Mansfield and Yeon, 1996).

3.2.2.3 Marketing and promoting products

MSMEs require reacting promptly to the growing marketing needs and innovations so as to withstand the increasing competition from large firms from within and outside. This sector requires better access to market amenities in order to sustain and further improve its involvement toward employment generation, output, and exports (Kaman et al., 2001; Swain and Pratihar, 2002; Hussain et al., 2011).

A published research has already highlighted that an enormous opportunity exists for MSMEs to accomplish their required financial goals by recognizing their presence and potential. Furthermore, it is revealed that since most of India's MSMEs, particularly the small scale industry, generate a large percentage of their profits from the local market; they still rely on conventional media like newspapers and telephone directories to contact their customers (Hall, 2002).

3.2.2.4 Manufacturing technology entirely new to firm

The manufacturing MSMEs in today's globalized economy are facing hard competition and rising demands for better quality services and products which are characterized by reliable deliveries, fast response time, and new and/or improved production methods (Stokes and Fitchew, 1997). In such a dynamic environment, development is considered as a main strategic factor for competitiveness of these manufacturing MSMEs. But technological development has been revealed to take many ways that reflect the numerous sources of production upon which it is based, i.e., one of the critical insights of current manufacturing theory is that firms hardly innovate only on the basis of internal resources, but they draw on skills, knowledge, technical solutions, equipment, and methods from outside the firm itself (Wilson, 1995).

Most of the manufacturing firms have complex relationships with customers, research institutes, suppliers, industry associations, and so on. This kind of interdependence has led to a broad set of models of development based on 'interactive learning' among firms and their wider environment (Bacon et al., 1996; Abereijo et al., 2009).

3.2.2.5 Financial strategies for utilization of funds

Developing an effective financial program that supports educational and training activities for enhancing organizational capabilities has become essential for highly innovative companies (Souitaris, 2002). Manufacturing industries in developed economies spend a considerable portion of their annual turnover on technology innovation. Encompassing an innovation budget is the main reason that differentiates innovative and non-innovative firms (Radas and Bozic, 2009)

Decisions regarding financial strategy are influenced by the company's external atmosphere and must be analyzed for potential threats and opportunities (Sirmon and Hitt, 2009). The proposed strategy must have room for the interests and requirements of company owners, customers, and management as much as possible (Abereijo et al., 2009; Hall, 2002). Financial strategy is subject to the overall corporate strategy, i.e., it includes investment strategy, profit distribution strategy, legal relations strategy, and financing strategy (Carmeli and Tishler, 2004).

3.2.2.6 Loans from bank for technology innovation

Financial statement lending is a transaction technology depending primarily on the strength of financial statements of a borrower. This technology has two requirements: first, the borrower should have proper information regarding financial statements as per broadly accepted standards of accounting. Second, the borrower must possess a strong financial situation to be reflected in these statements. The loan agreement that takes place from analysis of these statements may reveal various contracting elements such as personal and collateral guarantees. Under financial statement lending, the lender examines the usual future cash flow of the MSME which will act as the primary source of reimbursement. Guarantee matters are difficult to implement precisely, because MSMEs are unable to make good communication and cooperation with the banks. In addition, unlike other lending technologies, financial statement lending is reserved for comparatively informational transparent firms (Lado and Wilson, 1994; Pfeffer, 1998; Talukder and Quazi, 2010).

Small business credit scoring is based mainly on hard information about the firm and its entrepreneur. This is combined with data on the firm composed by the financial institutions and usually from commercial credit bureaus. The data are inserted into a loan performance forecast model, which produces a score or summary statistic for the loan. But in case of MSMEs, the banks unavoidably have careful loan behavior, and additionally the assessment for credit loans is extremely strict (Berger and Udell, 2006).

3.2.3 Organizational culture and climate

Literatures related to organizational climate looks at crucial issues involving the effect of social interactions inside the organization, while climate is contextually linked with situations relating to how employees feel, think, believe, and respond toward the organization. However, this can be a matter of manipulation by individuals who are in positions of power in the organization (Acikgoz and Gunsel, 2011). Organizational climate is defined as the 'present perceptions of people inside a work environment with respect to the noticeable (physical, social, and political) nature of the individual relationships that affect the completion of work within a specific organization'. Shim (2010) defined organizational climate as 'employees' mutual knowledge and opinion with others in their place of work'.

Culture in small firms is defined as less written records and directions. People unfamiliar with this kind of culture often mistake informality for lack of concern. A small firm owner casually requesting a subordinate to complete an activity may express unspecified consequences for non-performance than those of a registered letter of demand (Huang and Brown, 1999; Thornhill, 2006).

Management's role and involvement are primary differences between small and large companies. Generally, the entrepreneur of a small firm has greater attention and control of the firm through ownership. They can easily recognize all employees' weaknesses and strengths as they often perform all routine activities or processes by themselves. They understand customer desires and habits and know the customer's representatives. They are familiar with the relative weaknesses and strengths of the firm's products/services and also its competitors (Berg and Harral, 1998).

3.2.3.1 Motivation of employees

It is a basic responsibility of every manager to motivate his subordinates or to generate the 'will to work' among them. It should also be kept in mind that a worker might be immensely competent at doing some work, but nothing can be accomplished if he is not willing to work. Creation of a 'will to work' prospective is observed as motivation in easy and true sense of the term (Tidd, 2001).

The main purpose of motivation is to create an environment for employees to work with a sense of discipline, responsibility, and loyalty so that the objectives of an organization are achieved successfully (Nganga, 2011). Motivational techniques are used to encourage employee growth. Some managers attempt to motivate their employees by using rewards, formal authority, and punishments but motivation is much more complicated. It involves the concept of family, growth, learning, team work, salary, other benefits, and the like (Laforet, 2013).

Motivating people so as to make their best involvement in the achievement of organizational goals is one of the most significant concerns of a manager or management. Therefore, it becomes imperative for him to realize what motivates people to perform as they do (Nganga, 2011). Some human activities are random and consist of reflexes and emotions, the majority of it is goal directed in the manner that it is meant at the satisfaction of some desire. Since the desires of the organization and the employees are not always the same, the manager can better assimilate these two sets of desires by achieving an insight into the needs of his employees and then channelize these into the direction of organizational needs (Tirkey and Badugu, 2012).

3.2.3.2 Training of employees

Training beefs up the stock of technical resources within an organization and this occasionally has resulted in the formation of a pool of information that can be used in improving and modifying production processes in organizations. It has been found by Thornhill (2006) that innovative MSMEs are mainly those whose managers/owners were at one time apprentices in similar large firms. Knowledge acquired in the period of apprenticeship were then transferred to their firms and used to enhance the firms' performance, modifying the production products and processes. Also studies in African countries like Zimbabwe and Ghana (Nelson and Mwaura, 1997) have highlighted that the training characteristic of on-the-job training or learning by doing through apprenticeship has been helpful in building up the capabilities, local technological base, and also the development of innovations in small scale firms.

Training is one of the essential avenues through which technological knowledge and skills are transferred and innovative capabilities are built and improved. But training is usually of an informal nature in the MSMEs. Although public training institutions are instrumental in imparting knowledge and skills to technicians, artisans, and other middle level personnel in production, their impact has been comparably negligible in enhancing the innovation capabilities of those trained (Adner and Helfat, 2003; Hallier, 2009).

Apart from the unfavorable effects that liberalization had on formal training, the in-house informal training of apprentices was also affected. Due to the necessity for MSMEs to survive in the present liberalized market environment, several firms took methods to reduce cost. Unluckily, one of the areas where these firms had to cut down on expenses was on apprentice training. Even though the apprentice system is mainly 'geared to the transformation of traditional skills at reasonably low levels of technological sophistication' (Lall, 1992), it used to play an imperative role in providing the skilled labor to MSMEs. These developments in the apprenticeship training programs and education system in the country therefore had a positive impact on skill development and hence on technology innovation capabilities of the firm because a skilled work-force is essential for technology transformation in MSMEs.

3.2.3.3 Availability of skilled manpower

The quantum and quality of the technical personnel employed increase the possibility of an organization being engaged in technological research and development and its probability of success in undertaking innovation.

There exists a direct relationship between the quality and quantity of skilled workforce such as technicians, engineers, and highly skilled artisans in a firm and innovation since skilled employees bring about innovation through the learning-by-doing process (Scott et al., 1986).

For reacting to the changing market environment and lowering the cost structure, the organizations are required to use contingent labor (Matusik and Hill, 1998). These workers represent a variable rather than permanent cost for the firm and are engaged in production only when their particular skill set and/or productivity and knowledge is essential. Moreover, contingent labor can be released easily once their involvement is no longer required in an organization (Foote and Folta, 2002).

3.2.4 Government initiatives

Government policies affect the functioning of the MSMEs at a large scale. Providing external technological support to MSMEs and technology transfer rather than innovation and R&D assistance has been the key characteristic of government assistance programs for technology development in Indian MSMEs. Government policies look into MSME development and to evaluate performance with respect to their contributions toward entrepreneurship and employment generation (Pasadilla, 2010; Subrahmanya, 2005).

3.2.4.1 Government support in acquiring latest technology

India recognized the necessity for improving the competitive strength of small firms through technology development and modernization as early as in the 1950s with the setting up of:

(i) A network of Small Industries Service Institutes (SISIs) and Development
(ii) Commission for Small Scale Industry (DCSSI) and their extension centers
(iii) National Research Development Corporation (NRDC), and
(iv) National Small Industries Corporation (NSIC).

At this stage, government policy being emphasized promotion of modern small firms, it was realized that what was critical was to provide technical information, technical training or advice, and technology itself. Thus, SISIs along with DCSSI were set up to offer these services. NSIC came into existence to give machinery on hire purchase, apart from providing marketing support. NRDC was established with the aim of commercializing indigenous technologies developed in CSIR laboratories. In the process the issue of promoting the ability of small firms to carry out R&D and generate technological developments appears to be mistreated (Pasadilla, 2010).

In the 1990s, with the beginning of economic reforms and inherent shift in policy toward improving the competitiveness of the small scale sector, technology development was supposed to have greater consequences on these firms. But hardly any policy measure or scheme has been introduced to support the R&D and technology development activities of small scale industry. This is evident from the structure of institutions which presently provide technological support to small firms and policy measures initiated in the 1990s. Certainly, DSIR has introduced a scheme of incentives to encourage R&D in industry, which is applicable to small firms as well. But these incentives just intended at providing financial concessions more than anything else. Further, there is no clear scheme to encourage communication between small firms and research institutions for R&D and technology development (Subrahmanya, 2015).

3.2.4.2 Funds for R&D initiatives

The nature and level of R&D in small industries are evaluated with respect to the issues such as how many firms have carried out R&D, why R&D has been carried out in these firms, and what are the dimensions, objectives, sources, intensity, and the achievements of R&D. The role of external support, mainly that from government-promoted institutes, and inputs committed for R&D such as capital and personnel are also examined. The government has been following the path of globalization and liberalization since 1991 through de-licensing of industries, dismantling of regulations and controls for existing as well as new investments, radical reduction of tariff barriers for imports, elimination of constraints for foreign investment, and phasing out of quantitative limitations. This has increased local competition and exposed small industries to international competition. As a result, reacting to technological changes and fulfilling the expectations of customers have become crucial for survival and growth of a small firm (Subrahmanya, 2005).

Further, in order to assist the MSMEs in fully exploiting their potential by increasing their competitiveness to take on the challenges of hard competition as well as availing opportunities created by trade liberalization, the government in its NRDC scheme announced 'major promotional package' for MSME sector to give advantage in technological up-gradation, credit, and industrial infrastructure up-gradation (Jahanshahi et al., 2011).

3.3 CONCLUDING REMARKS

MSMEs are the fountain heads of Indian manufacturing and service sectors. The entrepreneurs of small firms in India are making progress in various industrial sectors such as manufacturing, food processing, textiles, garments, pharmaceutical, information technology, agro, retail, service sector, etc. In spite of the remarkable contribution of the MSME sector to the economy of the nation, concerned government departments, financial institutions, banks, and the corporate sector are not supporting them significantly. This is one of the main reasons for which small firms are competitive in both national and international markets. MSMEs face numerous problems like incomplete and inadequate market knowledge, non-availability of skilled labor at reasonable prices, lack of sufficient and timely finance, lack of appropriate and latest technology, unsuccessful marketing strategies, low production capability, lack of penetration into new markets, interaction with various government departments to resolve issues, and so on.

There have been many studies that establish the main factors contributing to the technological innovation initiatives of organizations especially MSMEs. Further, there have been very few empirical studies reported in literature that support the theoretical findings. With this backdrop, the present study is an attempt to focus on the empowering of the small firms. As discussed earlier, MSMEs significantly contribute toward economic development and growth in India; therefore it is pertinent to study and analyze the barriers of growth within small firms and identify technological innovation initiatives to improve manufacturing performance of the selected class of industry.

Chapter 4

Reliability analysis of technology innovation initiatives

4.1 INTRODUCTION

This chapter presents the results of a detailed survey conducted in Indian Micro, Small, and Medium Enterprises (MSMEs) covering various parts of the Northern region. The objective of the survey is to assess the status of major problems faced by small scale industry and at the same time to establish the relationship of various technology innovation initiatives (TIIs) with manufacturing performance parameters (MPPs) in order to achieve enhanced manufacturing performance. The results of the chapter reveal that Indian small firms have been reasonably successful in improving the manufacturing performance by implementing TIIs. The analysis of preliminary data has served as a useful input for the case studies and formulation of a strategic technology innovation program for small scale industry. Therefore, it can be concluded that Indian small firms must continue to make an earnest effort in their endeavor to realize enhanced performance through synergizing different 'technology innovation initiatives'.

4.2 ANALYSES OF PRELIMINARY DATA

From the enormous studies conducted by various researchers (Thampy, 2010; Subrahmanya, 2012; Laforet, 2013; James et al., 2014), it is found that *Entrepreneurial Capability, Technology Infrastructure Capability, Organizational Culture and Climate*, and *Government Initiatives* can help in improving the technology innovation in small firms. The following section deals with the evaluation of TIIs identified through an extensive literature review to ascertain their status in the selected class of industry.

4.2.1 Entrepreneurial capability (EC) issues

The performance of manufacturing organizations regarding EC issues has been shown in Table 4.1. Most of the organizations have generally scored

DOI: 10.1201/9781003272977-4

Table 4.1 Evaluation of EC Issues

S.No	Topic in the Aspect	No. of Responses (N)	No. of Companies Scoring				Total Points Scored (TPS)#	Percent Points Scored (PPS)	Central Tendency (TPS/N)
			1 (J)	2 (K)	3 (L)	4 (M)			
1	Good education level of entrepreneur (both general and business specific)	135	19	16	45	55	406	75.19	3.0
2	Ability to make effective decisions pertaining to business activities	135	28	24	39	44	369	68.33	2.7
3	Knowledge regarding various government schemes for MSMEs	135	25	20	41	49	384	71.11	2.8
4	Technical competencies of entrepreneur like competency in operating all machines, quality control tools, etc.	135	24	30	42	39	366	67.78	2.7
5	Entrepreneur training	135	29	31	35	40	356	65.93	2.6
6	Strategic decision making in identifying right kind of business and market	135	21	18	46	50	395	73.15	2.9
7	Tactic knowledge obtained through prior working experience	135	29	28	38	40	330	61.11	2.4
8	Alertness and formal business planning of different management areas like finance, H.R, logistics, etc.	135	40	41	29	25	309	57.22	2.3
9	Knowledge about various financial schemes and procedures to be followed to obtain loans, etc.	135	33	39	38	25	325	60.19	2.4
10	Ability to seize opportunities from market and using appropriate strategies in commercializing new product	135	42	45	25	23	299	55.37	2.2

(Continued)

Table 4.1 (Continued) Evaluation of EC Issues

S. No	Topic in the Aspect	No. of Responses (N)	No. of Companies Scoring				Total Points Scored (TPS)#	Percent Points Scored (PPS)	Central Tendency (TPS/N)
			1 (J)	2 (K)	3 (L)	4 (M)			
11	Awareness about new production technologies, machines, equipments, etc.	135	43	46	24	22	295	54.63	2.2
12	Strong emphasis on the development of new innovative products or improved products	135	33	25	38	39	353	65.37	2.6
13	Strong emphasis on R&D, technological leadership, and innovations	135	30	22	43	40	363	67.22	2.7
14	Ability to introduce new products services, techniques, and technologies	135	38	36	36	25	318	58.89	2.4
	Overall								2.6

{(1×K) + (2×K) + (3×L) + (4×M)}

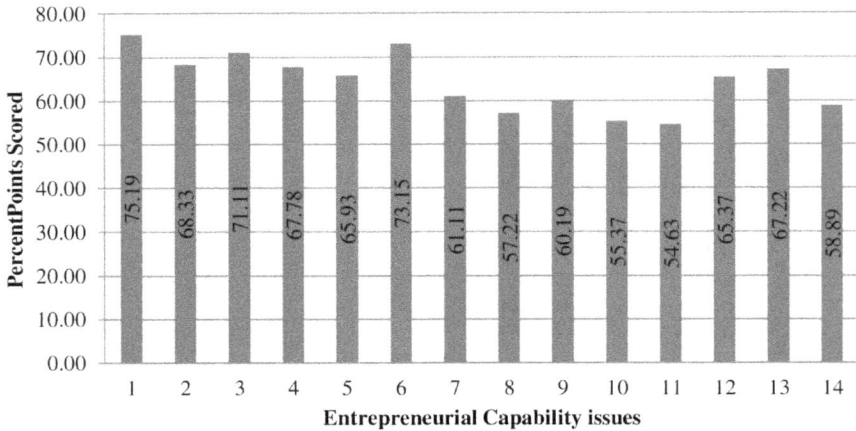

Figure 4.1 Issue-wise performance regarding EC issues.

high ratings (personal point scored) regarding education level and knowledge of entrepreneurs as indicated by the results.

The data shows that most of the organizations have a good education level of entrepreneurs (PPS = 75.19) which helps in strategic decision making to identify the right kind of business and market (PPS = 73.15). Also they have the ability to make effective decisions pertaining to business activities (PPS = 68.33).

At the same time, the entrepreneurs of small firms are not well aware about new production technologies, machines, equipments, etc. (PPS = 54.63) and also unable to introduce new products, services, techniques, and technologies in the organization (PPS = 58.89). Figure 4.1 depicts the overall performance of organizations regarding various EC issues.

4.2.2 Technology infrastructure capability (TIC) issues

Table 4.2 depicts the performance of manufacturing organizations regarding TIC issues. The analysis reveals that most of the organizations have technical knowledge and infrastructure to do business operations with information systems (e–purchasing, use of RFID and bar codes, etc.) as shown by the data (PPS = 75.19). Most of the organizations also have sufficient credit for meeting requirements of routine operations (PPS = 74.26). But on the other hand, the data of the survey reveals that small firms don't make specific financial strategies for effective utilization of available funds (PPS = 55.19).

Most of the organizations are unable to transform the results of R&D into products that meet market needs (PPS = 58.52) as well as acquire

Table 4.2 Evaluation of TIC Issues

S. No	Topic in the Aspect	No. of Responses (N)	No. of Companies Scoring				Total Points Scored (TPS)#	Percent Points Scored (PPS)	Central Tendency (TPS/N)
			1 (J)	2 (K)	3 (L)	4 (M)			
1	Appropriate raw materials at reasonable prices	135	15	10	52	58	423	78.33	3.1
2	Appropriate process technology infrastructure, i.e., state-of-the-art production machinery, equipments, tools, etc. to meet the changing market demands	135	24	23	42	46	380	70.37	2.8
3	Technical knowledge and infrastructure to do business operations with information systems (e-purchasing, use of RFID and bar codes, etc.)	135	19	16	45	55	406	75.19	3.0
4	Internet for marketing and promoting products	135	29	31	35	40	356	65.93	2.6
5	Implementation of new types of production processes	135	32	29	36	38	350	64.81	2.6
6	Acquiring manufacturing technology and skills entirely new to the firm	135	36	30	35	34	337	62.41	2.5
7	Organization's ability to transform the results of R&D into products that meet market needs	135	33	27	38	37	316	58.52	2.3
8	Softwares for designing and production-related tasks for product development	135	35	32	37	31	334	61.85	2.5
9	Financial budget, especially for R&D initiatives and new product development projects	135	28	33	39	35	351	65.00	2.6
10	Systematic budget system, i.e., earmarking of funds in advance for specific R&D initiatives and projects	135	32	31	39	33	343	63.52	2.5
11	Specific financial strategies for effective utilization of available funds	135	42	47	22	24	298	55.19	2.2
12	Sufficient credit for meeting requirements of routine operations	135	20	18	43	54	401	74.26	3.0
13	Getting loans from banks for technology up-gradation initiatives	135	31	27	37	40	356	65.93	2.6
	Overall								2.7

{(1×J) + (2×K) + (3×L) + (4×M)}

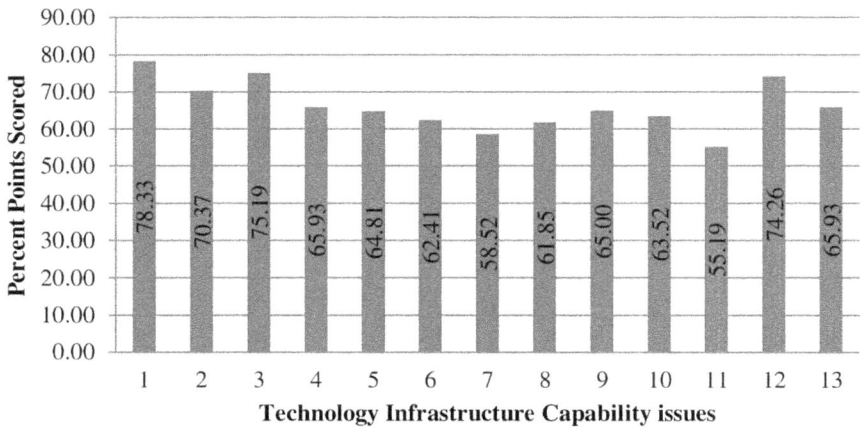

Figure 4.2 Issue-wise performance regarding TIC issues.

manufacturing technology and skills that are entirely new to the firm (PPS = 62.41). Figure 4.2 shows the overall performance of organizations regarding various TIC issues.

4.2.3 Organization culture and climate (OCC) issues

Table 4.3 depicts the performance of manufacturing organizations regarding OCC issues. The data shows that a reasonable number of organizations do not have skilled man power to increase competitiveness and suitable growths (PPS = 51.67) and at the same time, they do not provide adequate training to transfer knowledge and skills that are of requisite quality (PPS = 51.48). The extent of use of market and customer feedback into the innovation process is reasonable (PPS = 52.96) and most of the organizations do not have R&D personnel (PPS = 45.93). Figure 4.3 illustrates the overall performance of organizations regarding various organizational culture and climate issues.

The data shows that a reasonable number of organizations use formal reward structure to motivate employees (PPS = 57.41) and technological innovation initiatives are also supported by management (PPS = 67.41).

4.2.4 Government initiative (GI) issues

Table 4.4 depicts the performance of manufacturing organizations regarding GI issues. The analysis regarding government initiative issues reveals that getting cheap and reliable power supply is one of the major problems faced by MSMEs (PPS = 33.52). Good rail and road infrastructure also

Table 4.3 Evaluation of OCC Issues

S. No.	Topic in the Aspect	No. of Responses (N)	No. of Companies Scoring				Total Point Scored (TPS)#	Percent Point Scored (PPS)	Central Tendency (TPS/N)
			1 (J)	2 (K)	3 (L)	4 (M)			
1	Skilled manpower to increase competitiveness and sustainable growth	135	49	48	18	20	279	51.67	2.1
2	Intensity of R&D personnel (ratio between the number of full-time employees engaged in R&D and total number of employees in the organization)	135	58	52	14	11	248	45.93	1.8
3	Extent of training given to employees to transmit knowledge and skills of requisite quality	135	50	45	22	18	278	51.48	2.1
4	Formal reward structure to motivate employees (financial rewards, promotions, and carrier development opportunities)	135	40	42	26	27	310	57.41	2.3
5	Employee empowerment to take responsibility for technological innovation initiatives	135	31	28	37	39	354	65.56	2.6
6	Management support for technological innovation initiatives	135	28	28	36	43	364	67.41	2.7
7	Education level/qualification of production personnel	135	22	25	42	46	360	66.67	2.7
8	Extent of using market and customer feedback into the innovation processes	135	44	50	22	19	286	52.96	2.1
	Overall score								2.3

$\{(1 \times J) + (2 \times K) + (3 \times L) + (4 \times M)\}$

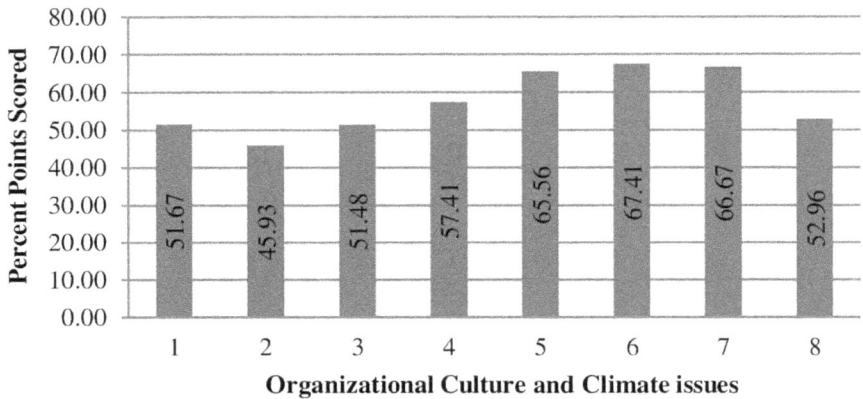

Figure 4.3 Issue-wise performance regarding OCC issues.

affects the MSME performance to a large extent (PPS = 48.15). Also the tax policies framed by government for MSMEs are not favorable for encouraging them to speed up technological and new product development projects (PPS = 44.63).

Government helps MSMEs in acquiring latest technology, quality certification, and marketing assistance up to certain extent (PPS = 56.67) as well as in allocating funds for R&D initiatives (PPs = 56.85). They are also provided with free or subsidized information regarding latest trends and technologies in relation to government regulations (PPS = 67.96). Figure 4.4 depicts the overall performance of organizations regarding various government initiative issues.

4.2.5 Manufacturing performance parameter (MPP) issues

Table 4.5 depicts the status of small firms regarding MPP issues. The analysis of survey reveals that TIIs have yielded considerably significant improvement in the manufacturing performance of organizations in terms of increase in cost/benefit ratio, improvement in product life cycle of products, reduction in the time to develop new products, ensuring increase in market share, and sales improvement through reduction in the cost of production and implementation of new processes.

It is evident from the survey that market share of a large number of organizations has been increased because of new products (PPS = 75.93) and there is improvement in product life cycle (PPS = 75.19) as a result of TIIs. There is a considerable improvement in sales due to new products as a percentage of total sales (PPS = 74.63). Figure 4.5 shows the overall performance of manufacturing organizations regarding MPP issues.

Table 4.4 Evaluation of GI issues:

S. No.	Topic in the Aspect	No. of Responses (N)	No. of Companies Scoring				Total Point Scored (TPS)#	Percent Point Scored (PPS)	Central Tendency (TPS/N)
			1 (J)	2 (K)	3 (L)	4 (M)			
1	Providing support in acquiring latest technology, quality certification, and marketing assistance	135	40	46	22	27	306	56.67	2.3
2	Providing training to employee at government institutes like tool rooms	135	31	29	37	38	352	65.19	2.6
3	Providing lab facilities for R&D initiatives at subsidized rates	135	21	18	46	50	395	73.15	2.9
4	Free or subsidized information regarding latest trends and technologies in relation to government regulations	135	27	26	40	42	367	67.96	2.7
5	Increase in important policies and measures to support innovation initiatives in MSMEs	135	29	27	33	46	366	67.78	2.7
6	Allocating funds for R&D initiatives in MSMEs	135	40	45	23	27	307	56.85	2.3
7	Tax policies for MSMES to encourage them to speed up technological and new product development projects	135	59	56	10	10	241	44.63	1.8
8	Providing cheap and reliable power supply to the MSMEs	135	60	54	11	10	181	33.52	1.3
	Providing good rail and road infrastructure in the region	135	54	52	14	15	260	48.15	1.9
	Overall score								2.3

{(1×J) + (2×K) + (3×L) + (4×M)}

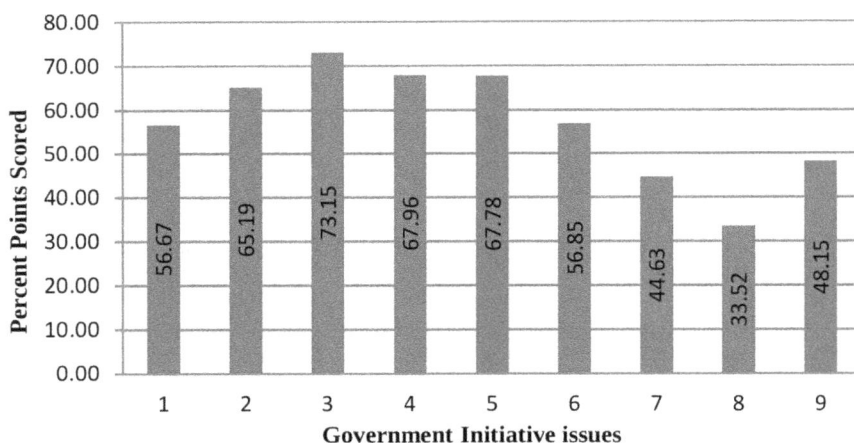

Figure 4.4 Issue-wise performance regarding GI issues.

4.3 RELATIONSHIP BETWEEN VARIOUS TIIs AND MPPs

A structural approach has been employed in this study for implementing the technology innovation (TI) program effectively to enhance manufacturing performance of organizations through TIIs. This underlines the significance of identifying the problems faced by MSMEs which affect their manufacturing performance. Based on the literature review and keeping in mind the problems faced by MSMEs, several measures are determined that can assist in implementing the TI program.

Therefore, the performance evaluation has been based on multiple inputs and multiple outputs and it can be thought of as a multi-criteria problem. In order to ascertain the benefits and contributions made by the TI program, it becomes imperative that various TIIs and MPPs be scrutinized carefully. In the present study, four input (I1, I2, I3, and I4) and three output (O1, O2, and O3) parameters have been identified as significant for analyzing the impact of the TI program toward achieving manufacturing performance enhancement in small firms as shown in Table 4.6. In order to evaluate the extent of deployment of various TIIs and MPPs as a result of these initiatives, a four point Likert Scale has been employed in this study as follows:

Technology innovation initiatives

(i) Not at all, (ii) To some extent, (iii) Reasonably well, (iv) To a great extent.

Table 4.5 Evaluation of MPP Issues

S. No	Topic in the Aspect	No. of Responses (N)	No. of Companies Scoring				Total Points Scored (TPS)#	Percent Points Scored (PPS)	Central Tendency (TPS/N)
			1 (J)	2 (K)	3 (L)	4 (M)			
1	Increase in cost/benefit ratio of company's products because of technology up-gradation initiatives as compared to competitor's products	135	20	18	46	51	398	73.70	2.9
2	Improvement in product life cycle of the products	135	19	15	47	54	406	75.19	3.0
3	Reduction in the time (from inception of idea to launch) to develop new products	135	20	17	46	52	400	74.07	3.0
4	Reduction in the cost of production	135	22	20	45	49	393	72.78	2.9
5	Increase in the proportion of new products as a percentage of total products over the last 3–5 years	135	26	28	40	41	366	67.78	2.7
6	Improvement in technical characteristics and features of existing product range	135	28	26	39	42	365	67.59	2.7
7	Implementation of new processes as a result of technology up-gradation initiatives	135	32	28	35	40	353	65.37	2.6
8	Extension in range of products in the last 3–5 years	135	42	37	27	29	271	50.19	2.0
9	Adaptation of the basic and key technologies	135	40	41	28	26	310	57.41	2.3
10	Mean sales profitability (increase in profit margins) over the last 3–5 years	135	46	39	23	27	301	55.74	2.2
11	Sales improvement due to new products as a percentage of total sales	135	19	17	46	53	403	74.63	3.0
12	Increase in market share because of new products	135	17	18	43	57	410	75.93	3.0
13	Penetration into new markets	135	32	38	30	35	338	62.59	2.5
	Overall								2.7

{(1×J) + (2×K) + (3×L) + (4×M)}

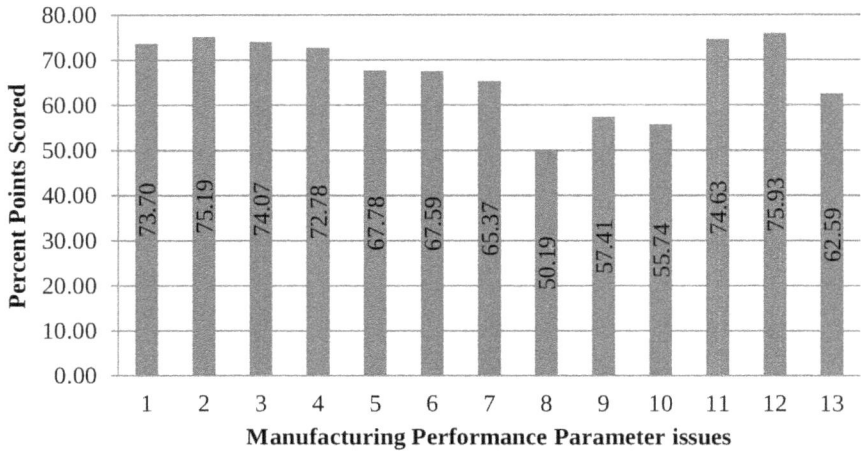

Figure 4.5 Issue-wise performance regarding MPP issues.

Table 4.6 Technology Innovation Initiatives and Manufacturing Performance
Parameters for Implementation of TI Program

Technology Innovation Initiatives (TIIs)	Manufacturing Performance Parameters (MPPs)
Entrepreneurial capabilities (I1)	Product performance (O1)
Technology infrastructure capability (I2)	Innovation performance (O2)
Organizational culture and climate (I3)	Sales performance (O3)
Government initiatives (I4)	

Manufacturing performance parameters

(i) Not at all, (ii) To some extent, (iii) Reasonably well, (iv) To a great
   extent, and
(i) <5%, (ii) 5–10%, (iii) 10–30%, (iv) >30%.

The study employs various statistical tools for taking out significant factors
contributing toward realization of manufacturing performance enhancements. For this purpose, various statistical tools like Cronbach's alpha,
Pearson's correlation coefficient, multiple regression analysis, 'T-test', and
canonical correlation have been employed to investigate the contribution
of TIIs toward realization of manufacturing performance enhancements in
small scale industry. Further, the following hypotheses have been formulated in the study:

Hypothesis related to TIIs and MPPs:

H1: There is a significant relationship between entrepreneurial capability and manufacturing performance parameters.

H2: There is a significant relationship between technology infrastructure capability and manufacturing performance parameters.

H3: There is a significant relationship between organizational culture and climate and manufacturing performance parameters.

H4: There is a significant relationship between government initiatives and manufacturing performance parameters.

4.4 CONTRIBUTION OF TIIS IN ACHIEVING MANUFACTURING PERFORMANCE ENHANCEMENT

The responses received from industrial surveys have been compiled and analyzed significantly to ascertain the performance of MSMEs regarding various TIIs. Moreover, the Cronbach's alpha for various categories of TIIs and MPPs has been evaluated to determine the reliability of input and output data collected through 'TI questionnaire'. The value of Cronbach's alpha in excess of 0.8 indicates high reliability of data for various input and output parameters as shown in Table 4.7.

All the input (I1–I4) and output (O1–O3) variables are also validated by discriminant validity analysis. It can be seen from Table 4.8 that the covariance within the group is higher than covariances outside the group (diagonal values represent covariance within the group). This further validates different input and output variables.

On the basis of responses received from the industry, an evaluation of relationship between various TIIs and MPPs has been made. Table 4.9 depicts the Pearson correlations among various TIIs and MPPs. The Pearson correlations have been worked out to determine the significant factors contributing toward success of the TI program in a firm. Only those pairs with Pearson correlation 'r' greater than or equal to 17% and statistically significant at 5% level of significance are considered to have a significant association.

Further, 'T-test' has been conducted for investigating and validating the correlations between TIIs and MPPs. The results of 'T-test' at 5%

Table 4.7 Cronbach's Alpha for Input and Output Parameters

Category	I1	I2	I3	I4	O1	O2	O3
Cronbach's alpha	0.92	0.90	0.82	0.89	0.88	0.93	0.92

Table 4.8 Covariance Values of Input and Output Parameters

Covariance	I1	I2	I3	I4	O1	O2	O3
I1	0.086	0.020	0.015	0.029	0.056	0.026	0.050
I2	0.020	0.115	0.029	0.056	0.026	0.050	0.081
I3	0.015	0.029	0.108	0.005	0.015	0.009	0.055
I4	0.029	0.056	0.005	0.146	0.085	0.016	0.030
O1	0.056	0.026	0.015	0.085	0.212	0.015	0.040
O2	0.026	0.050	0.009	0.016	0.015	0.215	0.003
O3	0.050	0.081	0.055	0.030	0.040	0.003	0.191

Table 4.9 Karl Pearson's Correlation and T-test Analysis between 'Technology Innovation Initiatives' and 'Manufacturing Performance Parameters'

Technology Innovation Initiatives (Input Parameters)		Manufacturing Performance Parameters (Output Parameters)		
		Product Performance (O1)	Innovation Performance (O2)	Sales Performance (O3)
Entrepreneurial capabilities (I1)	r	0.33*	0.20*	0.25*
	t (p)	4.03(0.0001)	2.35(0.0200)	2.98(0.0034)
Technology infrastructure (I2)	r	0.43*	0.17*	0.30*
	t (p)	5.49(0.0001)	1.99(0.0486)	3.63(0.0004)
Organizational culture and climate (I3)	r	0.04	0.08	0.23*
	t (p)	0.46(0.6450)	0.93 (0.3563)	2.73(0.0072)
Government initiatives (I4)	r	0.31*	0.01	0.44*
	t (p)	3.76(0.0002)	0.12(0.9083)	5.65(0.0001)

*Correlation significance at 0.05 level (two-tailed).

significance level have been depicted in Table 4.9. The 't' values can also be calculated by using the empirical expression:

$$t = \frac{r\sqrt{n-2}}{\sqrt{1-r2}} \geq t_{n-2} \quad \left(\text{from } t \text{ tables}\right) \tag{4.1}$$

where $n - 2$ are the degrees of freedom for a particular test and r is the Pearson correlation coefficient between a particular TII and a manufacturing performance parameter.

The results of 'T-test' (Table 4.9) reveals that various TIIs and MPPs are closely associated since the significance factor p has been coming out to be less than 0.01 in most of the cases. Moreover 't' value within confidence limits corresponding to $n - 2$ (=133) degrees of freedom at significance level of 5% works out to be 1.98 (from statistical 't' tables). The value obtained

Table 4.10 Multiple Regression Analysis between 'Technology Innovation Initiatives' and 'Manufacturing Performance Parameters'

Manufacturing Performance Parameters	TI Initiatives	t Value	p Value	Beta Value (β)	Significance Value (F)	R/R² Value
O1	I1	2.83	0.005	0.3395		
	I2	5.04	0.001	0.4117	8.451	0.454/0.206
	I4	2.82	0.005	0.2611		
O2	I1	1.71	0.089	0.2416	2.644	0.274/0.075
O3	I1	2.03	0.034	0.2478		
	I2	2.99	0.003	0.2475		
	I3	2.66	0.008	0.2750	13.808	0.546/0.298
	I4	1.87	0.053	0.1752		

from 't' table (1.98) is lower than 't' values obtained for most of the input–output combinations as revealed in Table 4.10. This further validates the significant correlation between various TIIs and MPPs. In order to investigate the contribution of TIIs for manufacturing performance enhancement, the significant correlation thus obtained as a result of Pearson's correlation and 'T-test' is validated through 'multiple regression analysis'. The results of 'multiple regression analysis' are depicted in Table 4.10 and various notations employed in the table include: regression coefficient (beta coefficient, β) and multiple correlation coefficients (R value). The results imply that there is a significant contribution of various TIIs with respective MPPs.

Finally the inter-relationship between various TIIs and MPPs has been validated through canonical correlation analysis. This analysis has been believed to be an extremely effective and appropriate multivariate statistical technique for assessing the inter-relationships between multiple inputs and outputs, when there is little prior knowledge regarding such relationships. Canonical correlation measures the bivariate correlation between linear composites of the predictor (TIIs) and criterion variables (MPPs).

The results of the canonical correlation analysis are represented in Table 4.11. Column 1 indicates strong and significant canonical correlation function (r = 0.750 at F statistics probability of 0.00) between the predictor set of TIIs and the criterion set of MPPs. The multivariate test statistics have been confirmed to be statistically significant (p < 0.001). The redundancy indices are 0.283 and 0.248 for the dependent and independent canonical variates, respectively.

The canonical loadings for independent variate (I1, I2, I3, and I4) range from 0.396 to 0.701, whereas canonical loadings for dependent variate (O1, O2, and O3) have also been found to be loaded up to 0.690. To assess the validity of the canonical loadings, stability runs were made by dropping one variable at a time and re-executing the canonical correlation analysis.

Table 4.11 Canonical Correlation Analysis between 'Technology Innovation Initiatives' and 'Manufacturing Performance Parameters' with Stability Analysis

	Results with All Variables	Results After Deletion of			
		I1	I2	I3	I4
Canonical correlation (R)	0.750	0.677	0.653	0.736	0.706
Canonical root (R²)	0.562	0.459	0.426	0.542	0.499
F statistics probability	0.000	0.000	0.000	0.000	0.000
Dependent vitiate canonical loadings					
O1	−0.690	−0.808	−0.821	−0.847	−0.715
O2	−0.025	−0.232	−0.103	−0.014	−0.036
O3	−0.602	−0.711	−0.700	−0.687	−0.821
Shared variance	0.399	0.404	0.391	0.397	0.396
Redundancy index	0.283	0.216	0.218	0.257	0.234
Independent vitiate canonical loadings					
I1	−0.701	—	−0.839	−0.717	−0.755
I2	−0.682	−0.774	—	−0.683	−0.715
I3	−0.396	−0.439	−0.421	—	−0.457
I4	−0.584	−0.660	−0.664	−0.612	—
Shared variance	0.363	0.409	0.441	0.452	0.430
Redundancy index	0.248	0.224	0.238	0.281	0.250

Canonical loadings measure the correlation between TIIs and MPPs and their respective canonical variates, and these are similar in interpretation to factor loadings. Columns 3–6 in Table 4.11 show the results of these stability runs corresponding to deletion of the criterion variables I1, I2, I3, and I4 respectively. As can be seen in column 3–6 of Table 4.11, the results indicate the stability of canonical loadings.

The critical examination of Pearson's correlation and 'T-test' results explicitly depicts that adaptation of strategic TIIs (I1, I2, I3, and I4) has considerable impact on realization of overall manufacturing performance enhancement in the organizations. The results reveal that entrepreneurial capability, technology infrastructure capability, and government initiatives (I1, I2 and I4) can strategically contribute toward recognition of manufacturing performance enhancements by improving the product performance and sales performance.

4.4.1 Validation of hypotheses

The canonical correlation function of r = 0.750 at F statistics probability of 0.00 reveals a significant overall association between the criterion set of TIIs and the predictor set of MPPs. Thus, the hypotheses are validated in the present context. Moreover, the Pearson's correlations show that there exists a significant association between various TIIs and strategic MPPs. The critical examination of the results of multiple regression analysis shows that entrepreneurial capability (I1) issues are considerably associated with all the manufacturing performance parameters (O1, O2, and O3) in small scale industry. Good education level of entrepreneurs ensures quick and right decisions, and the ability of entrepreneurs to identify the right kind of business and market is important for technology innovation activities in small firms. Also, knowledge regarding various government schemes for MSMEs and work experience of entrepreneurs play a vital role in technology innovation process.

Technology infrastructure capability (I2) issues are the second main input parameter for success of the TI program in the manufacturing enterprises. Results of the survey show that product performance (O1) and sales performance (O3) are highly affected by technology infrastructure capability issues. In the era of globalization and increasing competition, use of the internet for marketing and promoting products is becoming necessary. In this age where market demand changes almost every day, state-of-the-art production machinery, equipment, tools, etc. facilitate organizations to meet the changing market demand. This reveals the significance of an effective information system for small scale industry to provide information for managers on the routine operations of the enterprise.

Organizational culture and climate (I3) issues are closely associated with sales performance (O3) as indicated by the results of the study. This implies that organizational culture and climate contributes toward optimizing cost of production in the industry by effective utilization of R&D activities which helps MSMEs in technology innovation. Training given to employees to transmit knowledge and skills of requisite quality also has a significant effect on technology innovation process.

Government initiative (I2) issues are the last main input parameter for success of the TI program in the manufacturing enterprises. The critical examination of results reveals that government initiative (I2) issues are significantly linked with product performance (O1) and sales performance (O3) in the industry. Government initiatives facilitate small firms in acquiring quality certification and latest technology and provide subsidized information regarding latest trends and technologies. Government contributes toward progress of small scale industry by providing lab facilities for R&D initiative at subsidized rates and regulation of tax policies to speed up their technological and new product development projects. Therefore, it can be said that successful implementation of various TIIs significantly contribute toward enhanced manufacturing performance of small scale industry.

4.5 CONCLUDING REMARKS

The research highlights the major problems faced by MSMEs in the era of globalization. It also evaluates the contributions of various TIIs in the

Indian small scale manufacturing industry for acquiring strategic benefits in the competitive environment. The empirical analysis has been carried out in this study to examine the role of TIIs in achieving significant manufacturing performance improvements in the manufacturing organizations. For this purpose, various TIIs and MPPs have been established in the research. Results of the study reveal that TIIs such as entrepreneurial capability, technology infrastructure capability, and government initiatives are far more important in affecting manufacturing performance of small firms. This validates the extremely high potential of TIIs in realizing overall organizational growth.

The relationships indicate the effectiveness of considering manufacturing performance as a multi-dimensional concept. The findings suggest that effective implementation of the TI program significantly contributes toward realization of strategic manufacturing performance improvements for competing in the highly dynamic global marketplace. Small firms must be aware of the existing capabilities within the sector to be able to manage the strategic TIIs effectively to compete in the global marketplace.

Therefore, for successful implementation of the TI program in the industry, it becomes compulsory for the organizations to understand the performance and interaction of different facets of technology innovation, so that the concept can fulfill its true potential.

Chapter 5

Multi-criteria decision-making techniques

5.1 INTRODUCTION

This chapter presents a synthesis of learning issues, outcomes of survey and case studies, and their utilization through a qualitative model to evolve a technology innovation program for small scale industry.

Qualitative modeling used in this study involved deriving expert opinion and using this along with findings of previous phases (literature review, survey, and case studies) in a structured manner. For this purpose, experts were invited to participate in the exercise. The panel of experts was drawn from the participating industry and academic institutes. The detailed findings of previous phases were shared with the experts.

5.2 EVALUATION OF TIIS USING FUZZY-BASED MODEL

Nowadays all manufacturers are trying to implement new methods like multi-criteria decision making (MCDM) models such as analytical hierarchy process (AHP), piecewise-affine (PWA), a fuzzy-based (FB) model, etc. for effective utilization of production processes. In this study, an attempt has been made to show the synergistic suitability of transfusion of TI for the Indian small scale manufacturing industry. For this study, the most relevant factors affecting these initiatives, that is product performance and sales performance, have been considered, and further these factors have been simulated by the data given by experts in these fields using Fuzzy Logic Toolbox of MATLAB. It provides the steps for designing fuzzy inference systems using graphical tools, and a Simulink block for analyzing, designing, and simulating systems based on fuzzy logic.

The toolbox helps in building models of complex system behaviors using simple logic rules, and then implementing those rules in a fuzzy inference system. The study focuses on identifying the suitability of above said factors by the fuzzy-based simulation (FBS) model. These factors had been taken after considering the view points of TI coordinators from different manufacturing units.

DOI: 10.1201/9781003272977-5

A suitable TI method is expressed by the following equation:

$$\text{Suitable TI method} = f[\text{Product Performance, Sales Performance}] \quad (5.1)$$

Therefore the above equation is further optimized with the use of fuzzy logic. In recent years, the number and variety of applications of fuzzy logic have increased significantly. The result proves that technological innovation initiatives are necessary to enhance manufacturing performance of small firms.

5.3 FUZZY INFERENCE SYSTEMS (FIS)

Fuzzy inference is the process of formulating the mapping from a given input to an output using fuzzy logic. The mapping then provides a basis from which decisions can be made, or patterns discerned. Fuzzy inference systems have been successfully applied in fields such as automatic control, data classification, decision analysis, expert systems, and computer vision. Because of its multidisciplinary nature, fuzzy inference systems are associated with a number of names, such as fuzzy-rule-based systems, fuzzy expert systems, fuzzy modeling, fuzzy associative memory, fuzzy logic controllers, and simply (and ambiguously) fuzzy systems. The flowchart depicting the procedure of FIS used in the present study is shown in Figure 5.1. There are two types of fuzzy inference systems that can be implemented in the Fuzzy Logic Toolbox: Mamdani-type and Sugeno-type. Mamdani-type inference, as defined for the toolbox, expects the output membership functions to be fuzzy sets.

After the aggregation process, there is a fuzzy set for each output variable that needs defuzzification. It is possible, and in many cases much more efficient, to use a single spike as the output membership functions rather than a distributed fuzzy set. This type of output is sometimes known as a singleton output membership function, and it can be thought of as a pre-defuzzified fuzzy set. Sugeno-type systems can be used to model any inference system in which the output membership functions are either linear or constant. Fuzzy inference process comprises five parts: fuzzification of the input variables, application of the fuzzy operator (AND or OR) in the antecedent, implication from the antecedent to the consequent, aggregation of the consequents across the rules, and defuzzification.

5.3.1 Fuzzification

The first step is to take the inputs and to determine the degree to which they belong to each of the appropriate fuzzy sets via membership functions. In Fuzzy Logic Toolbox software, the input is always a numerical value limited to the input variable and the output is a fuzzy degree of membership in the qualifying linguistic set (always the interval between 0 and 1).

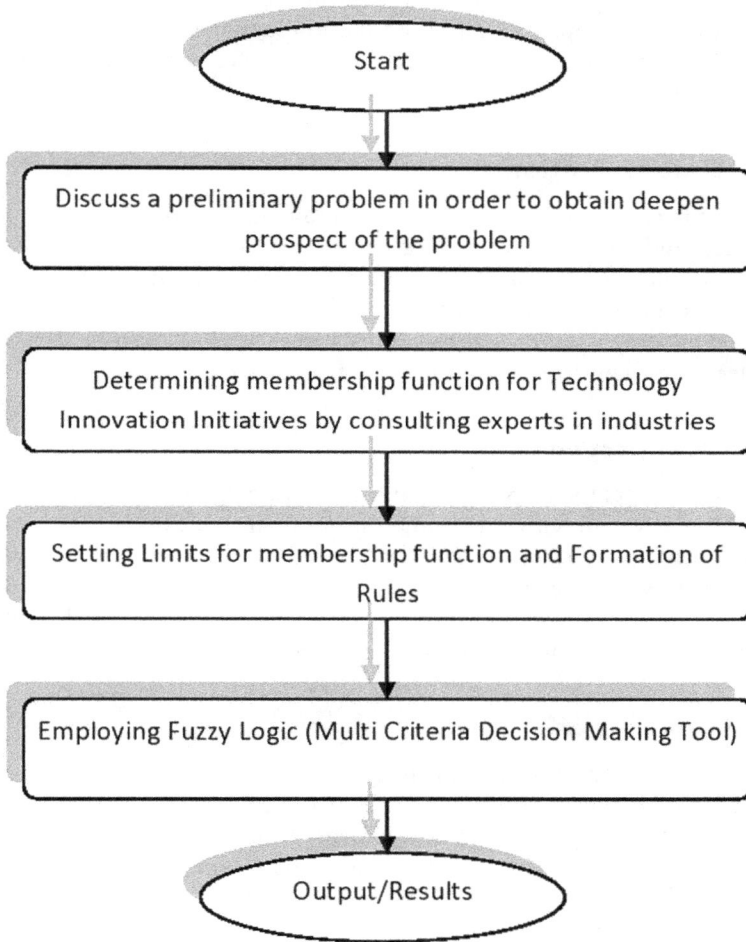

Figure 5.1 FIS procedure used in the present study.

5.3.2 Rule evaluation

The FIS develops appropriate rules and on the basis of the rules the decision is made. This is principally established on the concepts of the fuzzy set theory, fuzzy IF–THEN rules, and fuzzy reasoning. FIS uses 'IF … THEN …' statements, and the connectors existent in the rule statement are 'OR' or 'AND' to create the essential decision rules. The basic FIS can accept either fuzzy inputs or crisp inputs, but the outputs it provides are virtually all the time fuzzy sets. When the FIS is employed as a controller, it is needed to have a crisp output. Hence, in the present study the rules are formed with the expert knowledge, feedback, and guidance given by experts in the manufacturing industries and are further refined with experienced persons in the field of operation and production management, and are further refined following real life application and appraisal which will either confirm them or require them to be modified.

5.3.3 Defuzzification

The input for the defuzzification process is a fuzzy set (the aggregate output fuzzy set) and the output is a single number. As much as fuzziness helps the rule evaluation during the intermediate steps, the final desired output for each variable is generally a single number. However, the aggregate of a fuzzy set encompasses a range of output values, and so must be defuzzified in order to resolve a single output value from the set.

5.3.3.1 Fuzzification of TI

Figure 5.2 depicts the empirical transfer function of TI from equation (1) as a fuzzy logic system with inputs and output being fuzzified using appropriate membership functions. Here the inputs are the product performance and sales performance. The output is the result whose value shows whether to accept, under consider, or reject the selection of TI measures. The following sections narrate each component of the system.

5.3.3.2 Product performance

The product performance is measured by the cost of production or the cost/benefit ratio of the product, etc. By consulting various experts from the industries, the fuzzy set rules defined for product performance were set as follows: if the product performance is less or more than –8 to 8% of required value, then the system is considered as rejected or accepted. If it lies between 5 and 8%, it is considered as very high or very low. If the product performance is between 2 and 5% less or greater than the required value, it is considered as low or high. If the product performance is 2% less or greater than the required value, then it is considered as optimum as

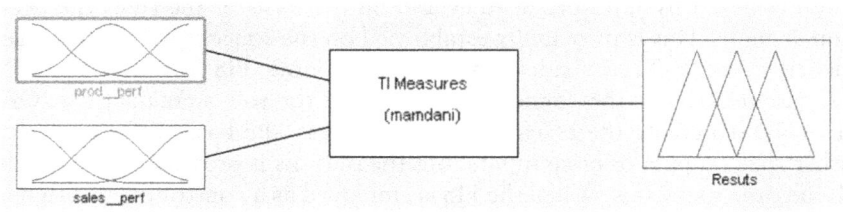

Figure 5.2 Empirical transfer function of TI.

Table 5.1 Range for Product Performance Measurement

Fuzzy	Linguistic Term	Range
I	Accept system	More than −8%
2	Very low	From −8 to −5
3	Low	From −5 to −2 %
4	Optimum	From −2 to 2 %
5	High	From 2 to 5 %
6	Very high	More than 5 to 8
7	Reject system	More than 8%

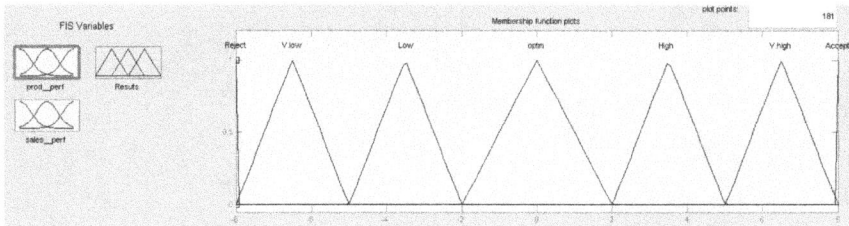

Figure 5.3 Transfer function in fuzzy format of product performance.

shown in Table 5.1. The transfer function in fuzzy format is represented in Figure 5.3.

5.3.3.3 Sales performance

The sales performance is associated with mean sales profitability or increase in market share of the company. By consulting various experts from the industries the fuzzy set rules defined for sales performance are: if the sales performance is less or more than 10% as compared to the previous data, then it is considered as rejected or accepted. If the sale is 8–10% less or greater, then it is considered as very low or very high. If the sale is 5–8% less or greater, then it is considered as low or high. If the sales performance is 5% less or greater than the required value, then it is considered as optimum as shown in Table 5.2. The transfer function in fuzzy format is shown in Figure 5.4.

5.3.4 Result: checking the suitability of TI measures

The result is used to decide whether to select the TI measures or not. The result value lying between 0 and 3 shows that TI measures should be rejected, between 3 and 6 represents that considered measures are poor (Under Consideration). Result value from 6 to 8 is considered as acceptable

Table 5.2 Range for Sales Performance Measurement

Fuzzy	Linguistic Term	Range
1	Accept system	More than −10%
2	Very low	From −10 to −8
3	Low	From −8 to −5 %
4	Optimum	From −5 to 5 %
5	High	From 5 to 8 %
6	Very high	More than 8 to 10
7	Reject system	More than 10%

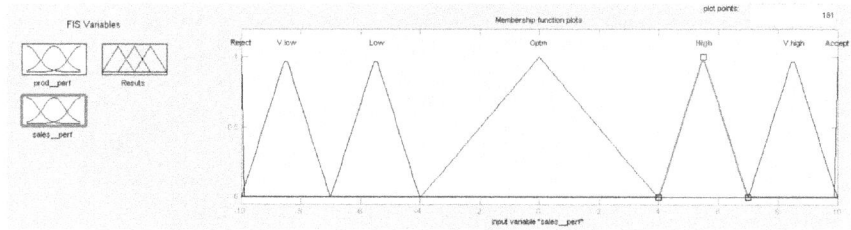

Figure 5.4 Transfer function in fuzzy format of sales performance.

Table 5.3 Range for TI-Result Measurement

Fuzzy	Linguistic Term	Range
1	Reject	0–3
2	Under consideration	3–6
3	Acceptable	6–8
4	Optimum	8–10

and from 8 to 10 is considered as Optimum as shown in Table 5.3. The transfer function in fuzzy format is shown in Figure 5.5.

5.3.5 Fuzzy evaluation rules and solution

Table 5.4 demonstrates the 25 rules following the format 'if then' (results) corresponding to the combinations of input conditions. A single fuzzy if–then rule assumes the form 'if x is A then y is B' where A and B are linguistic values defined by fuzzy sets. The if-part of the rule 'x is A' is called the premise, while the then-part of the rule 'y is B' is called the conclusion. For example 'If the Product Performance is very high' and 'Sales Performance is optimum' then the result is 'system is Acceptable'. The rules are determined through expert knowledge and are further refined following real life

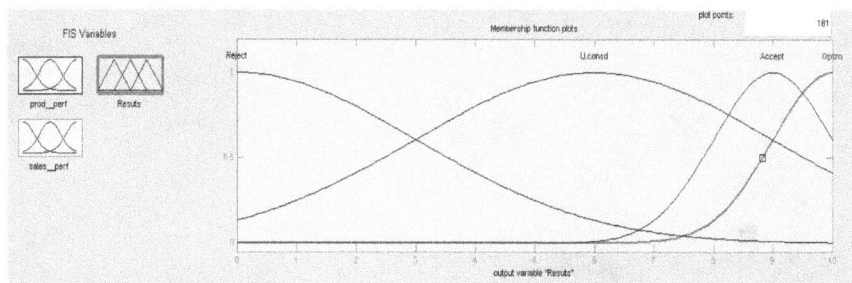

Figure 5.5 Transfer function in fuzzy format of TI-result.

Table 5.4 Demonstrating Fuzzy Rules for TI-Result

Sales_Pr \ Prod_Pr	Very Low	Low	Optimum	High	Very High
Very low	Reject	Reject	Reject	Reject	Reject
Low	Reject	Reject	Reject	Reject	Accept
Optimum	Reject	Reject	Accept	Accept	Accept
High	Reject	Reject	Accept	Accept	Accept
Very high	Reject	Reject	Accept	Accept	Accept

application and appraisal which will either confirm them or require them to be modified, as shown in Figures 5.6 and 5.7.

A continuum of fuzzy solutions for equation (1) is presented in Figure 6.8 using the rule viewer of the fuzzy tool box of MATLAB. The rule viewer displays a roadmap of the whole fuzzy inference process and it is based on the fuzzy inference diagram. The rule viewer allows interpreting the entire fuzzy inference process at once. It also shows how the shape of certain membership functions influences the overall result as it plots every part of every rule. Each rule is a row of plots, and each column is a variable. The rule numbers are displayed on the left of each row. By clicking on a rule number, the rule in the status line can be viewed. The two inputs can be set within the upper and lower specification limits and the output response is calculated as a score that can be translated into linguistic terms. In this instance the order output of 8.52 indicates 'Acceptable system' linguistically from Table 5.4. The rule viewer shows in detail one calculation at a time, and in this sense, it presents a sort of micro view of the fuzzy inference system.

5.3.5.1 Interpretations and conclusions

Technological innovation is a key factor in a firm's competitiveness. Technological innovation is unavoidable for firms which want to develop

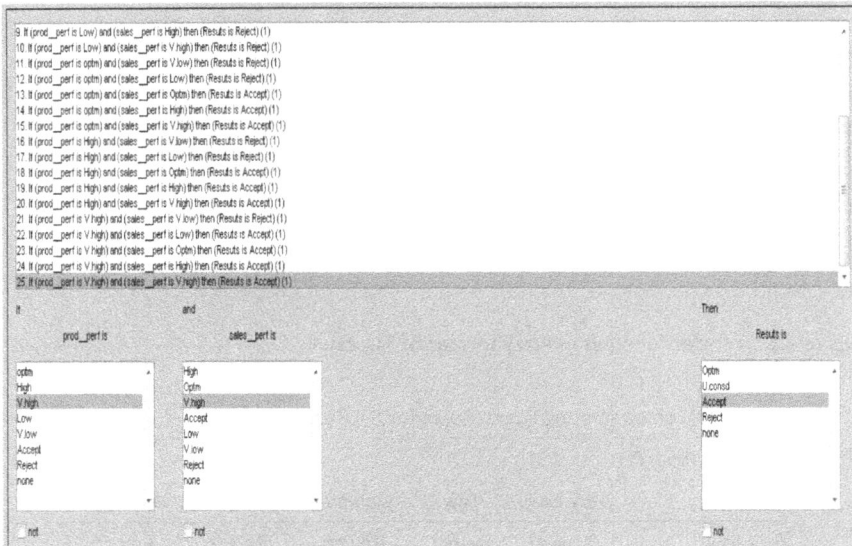

9. If (prod__pert is Low) and (sales__pert is High) then (Results is Reject) (1)
10. If (prod__pert is Low) and (sales__pert is V.high) then (Results is Reject) (1)
11. If (prod__pert is optm) and (sales__pert is V.low) then (Results is Reject) (1)
12. If (prod__pert is optm) and (sales__pert is Low) then (Results is Reject) (1)
13. If (prod__pert is optm) and (sales__pert is Optm) then (Results is Accept) (1)
14. If (prod__pert is optm) and (sales__pert is High) then (Results is Accept) (1)
15. If (prod__pert is optm) and (sales__pert is V.high) then (Results is Accept) (1)
16. If (prod__pert is High) and (sales__pert is V.low) then (Results is Reject) (1)
17. If (prod__pert is High) and (sales__pert is Low) then (Results is Reject) (1)
18. If (prod__pert is High) and (sales__pert is Optm) then (Results is Accept) (1)
19. If (prod__pert is High) and (sales__pert is High) then (Results is Accept) (1)
20. If (prod__pert is High) and (sales__pert is V.high) then (Results is Accept) (1)
21. If (prod__pert is V.high) and (sales__pert is V.low) then (Results is Reject) (1)
22. If (prod__pert is V.high) and (sales__pert is Low) then (Results is Accept) (1)
23. If (prod__pert is V.high) and (sales__pert is Optm) then (Results is Accept) (1)
24. If (prod__pert is V.high) and (sales__pert is High) then (Results is Accept) (1)
25. If (prod__pert is V.high) and (sales__pert is V.high) then (Results is Accept) (1)

Figure 5.6 Fuzzy set rules for TI-result.

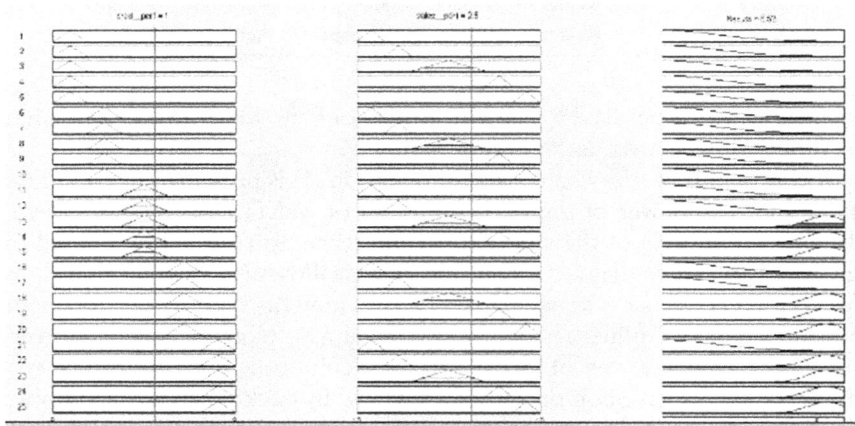

Figure 5.7 Rule viewer for TI-result.

and maintain a competitive advantage and/or gain entry in to new markets. There is substantial evidence to show that a number of Micro, Small, and Medium Enterprises (MSMEs) in a wide variety of sectors do engage in technological innovations, and that these innovations are likely to be an important determinant of their success. Therefore, in this study an assessment of the manufacturing industries has been done based on technology innovation initiatives using Fuzzy Logic Toolbox of MATLAB. Most important factors under TI, i.e., product performance and sales performance, have been considered as membership functions for fuzzy input as discussed by experts in this field. Also the expert opinion has been taken to formulate the fuzzy 'if then' rules.

Fuzzy interference system and the fuzzification process of TI approach have been formed during run time by assigning appropriate membership functions to the required approach. The result shows that synergistic effect of implementing TI measures efficiently improves manufacturing performance of an organization and this has been shown with the help of fuzzy rule viewer. Also to analyze the performance of fuzzy system, three dimensional plots output has been used with the help of surface view tool of Fuzzy Toolbox of MATLAB.

5.4 ANALYTIC HIERARCHY PROCESS (AHP)

The foundation of the analytic hierarchy process (AHP) is a set of axioms that carefully delimits the scope of the problem environment. The analytic hierarchy process (AHP) is widely used for tackling multi-criteria decision-making (MCDM) problems (Saaty, 1980). It is based on the well-defined mathematical structure of consistent matrices and their associated right-eigenvector's ability to generate true or approximate weights (Saaty 1980, 1994). It can be considered to be both a descriptive and prescriptive model of decision making. Its validity is based on the many hundreds (now thousands) of actual applications in which the AHP results were accepted and used by the cognizant decision makers (DMs) (Saaty, 1994). Thus AHP is the most widely used decision-making approach in the world today.

The AHP methodology compares criteria, or alternatives with respect to a criterion, in a natural, pair-wise mode. To do so, the AHP uses a fundamental scale of absolute numbers that has been proven in practice and validated by physical and decision problem experiments. The fundamental scale has been shown to be a scale that captures individual preferences with respect to quantitative and qualitative attributes just as well or better than other scales (Saaty 1980, 1994). In the late 1960s, Thomas Saaty, one of the pioneers of operations research, and author of the first *Mathematical Methods of Operations Research* textbook, developed the three primary functions of AHP. This primary function of AHP helps in understanding why it is such a general methodology with such a wide variety of applications. He sought a simple way to deal with complexity, simple enough so that lay people with no formal training could understand and participate. He found one thing common in numerous examples of the ways humans had dealt with complexity over the ages – that was the hierarchical structuring of complexity into homogeneous clusters of factors. Further he stated that any hierarchically structured methodology (like AHP) must use ratio scale priorities for elements above the lowest level of the hierarchy. This is necessary because the priorities (or weights) of the elements at any level of

Table 5.5 Basic Methodology for AHP

Intelligence phase	
Discuss a preliminary problem in order to:	obtain an enriched, consensual view of the problem.
Design phase	
Discuss an initial list of alternatives in order to:	obtain a revised list of alternatives; obtain an initial set of objectives/criteria.
Discuss an initial set of objectives in order to:	obtain a revised set of objectives/criteria.
Choice phase	
Structure one or more AHP models in order to:	obtain common (group) expert choice model(s) along with judgments.

the hierarchy are determined by multiplying the priorities of the elements in that level by the priorities of the parent element.

Because complex, crucial decision situations, or forecasts, or resource allocations often involve too many dimensions for humans to synthesize intuitively, we need a way to synthesize over many dimensions. Basic methodology to be used in AHP mainly consists of three phases as given by Saaty (1980): intelligence phase, design phase, and choice phase, which have been briefly explained in Table 5.5.

5.4.1 Describing model structure: the sub-objectives for decision-making

A thorough analysis of the problem is required for the identification of the important attributes (sub-objectives) involved. For current study, the selection of attributes has been determined through literature survey and discussions which were held with experts during industrial visits. Although there can be many factors which can affect decision-making in TI. We selected the factors given in Table 5.6.

Table 5.6 Description of Attributes

Attribute	Abbreviation
Entrepreneurial capability	EC
Technology infrastructure capability	TIC
Organizational culture and climate	OCC
Government initiatives	GI
Product performance	PP
Innovation performance	IP
Sales performance	SP

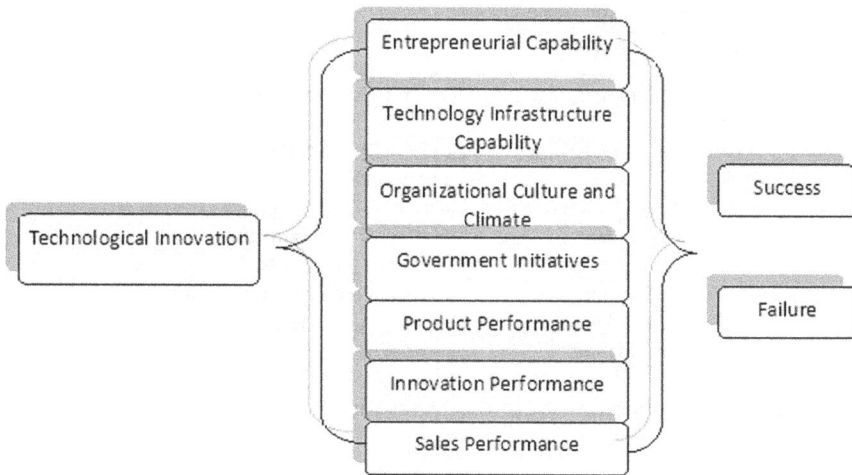

Figure 5.8 AHP model formulated.

5.4.2 Hierarchy formulated

The hierarchy for the decision making in AHP is formulated by breaking down the current problem statement into a hierarchy (levels) of decision elements and shown in Figure 5.8. In this we have taken a goal, seven attributes (EC, TIC, OCC, GI, PP, IP, and SP) and two alternatives (success and failure).

5.4.3 Scale used for pair-wise comparison of attributes

Pair-wise comparison is a key step in an AHP model to determine priority weights of factors and provide a rating for alternatives based on qualitative factors. The AHP focuses on two factors at a time and their relation to each other, so decision making will be more comfortable as to offer relative (rather than absolute) preference information.

The relative importance of each factor is rated by a measurement scale to provide numerical judgments corresponding to verbal judgments. The instrument used in this research is a discrete scale, from 1 to 9, with 1 representing the equal importance of two factors and 9 being the highest possible importance of one factor over another, as shown in Table 5.7.

5.4.4 Pair-wise comparison of different attributes

In this comparison, the importance of ith sub-objective is compared with j^{th} sub-objective is calculated. For this depending upon the number of

Table 5.7 Comparison Scale Used

Intensity	Definition	Explanation
1	Equal importance	Two factors contribute equally to the objective
3	Moderately more important	Experience and judgment favor one factor over another
5	Strongly more important	Experience and judgment strongly favor one factor over another
7	Very strongly more important	A factor is strongly favored and its dominance demonstrated in practice
9	Extremely more important	The evidence of favoring one factor over another is of the highest possible order of affirmation
2, 4, 6, 8	Intermediate values when compromise is needed	

attributes 7 (in our case), a 7 × 7 matrix was formed and following procedure was used to fill 7 × 7 matrix.

1. The diagonal elements of the matrix are always 1.
2. Upper triangular matrix was filled from the data available through companies.
3. To fill the lower triangular matrix, we use the reciprocal values of the upper diagonal, i.e., if aij is the element of row ith and column jth of the matrix, then the lower diagonal is filled using this formula aji = 1/aij. Thus, the pair-wise comparison matrix for different attributes is shown in Table 5.8.

5.4.5 Normalization of comparison matrix

Having a comparison matrix, the next step is to compute priority vector, which is the normalized eigenvector of the matrix. For this division was

Table 5.8 Pair-wise comparison matrix

Attribute	EC	TIC	OCC	GI	PP	IP	SP
EC	1	2	4	2	5	7	4
TIC	0.50	1	3	2	4	2	6
OCC	0.25	0.33	1	0.50	2	2	4
GI	0.50	0.50	2	1	5	3	4
PP	0.20	0.25	0.50	0.20	1	0.33	3
IP	0.14	0.50	0.50	0.33	3	1	2
SP	0.25	0.16	0.25	0.25	0.33	0.50	1
Sum	2.84	4.74	11.25	6.28	20.33	15.83	24

Table 5.9 Normalized Matrix

Attribute	EC	TIC	OCC	GI	PP	IP	SP
EC	0.352	0.422	0.355	0.318	0.246	0.442	0.167
TIC	0.176	0.211	0.267	0.318	0.197	0.126	0.250
OCC	0.088	0.070	0.089	0.080	0.098	0.126	0.167
GI	0.176	0.105	0.178	0.159	0.246	0.189	0.167
PP	0.070	0.053	0.044	0.032	0.049	0.021	0.125
IP	0.049	0.105	0.044	0.053	0.147	0.063	0.083
SP	0.088	0.034	0.022	0.040	0.016	0.031	0.042
Normalized sum	1	1	1	1	1	1	1

done of each entry in column by the sum of entries in column to get value of normalized matrix. Thus, the sum of each column is 1 as shown in Table 5.9. The normalized value r_{ij} is calculated as:

$$r_{ij} = a_{ij} / \sum_{i=1}^{n} a_{ij} \tag{5.2}$$

Further, the approximate priority weight (W_1, W_2, ... W_j) for each attribute is obtained as shown in Table 5.10.

$$W_j = 1/n \times \sum_{i=1}^{n} a_{ij} \tag{5.3}$$

5.4.6 Checking for consistency

Saaty (1980) proved that for consistent reciprocal matrix, the largest eigenvalue is equal to the size of comparison matrix, and then he gave a measure of consistency, called consistency index as deviation or degree of consistency. Considering this relative weight, which would also present the eigenvalues of criteria, should verify:

$$A \times W_i = \lambda_{max} \times W_i \quad i = 1; 2; ...; n \tag{5.4}$$

Table 5.10 Normalized Matrix Along with Priority Weights

Attribute	EC	TIC	OCC	GI	PP	IP	SP	Weight
EC	0.352	0.422	0.355	0.318	0.246	0.442	0.167	0.329
TIC	0.176	0.211	0.267	0.318	0.197	0.126	0.250	0.221
OCC	0.088	0.070	0.089	0.080	0.098	0.126	0.167	0.102
GI	0.176	0.105	0.178	0.159	0.246	0.189	0.167	0.174
PP	0.070	0.053	0.044	0.032	0.049	0.021	0.125	0.056
IP	0.049	0.105	0.044	0.053	0.147	0.063	0.083	0.078
SP	0.088	0.034	0.022	0.040	0.016	0.031	0.042	0.039

Table 5.11 Random Consistency Index (RI)

n	1	2	3	4	5	6	7	8	9	10
RI	0	0	0.58	0.9	1.12	1.24	1.32	1.41	1.45	1.49

where A represents the pair-wise comparison decision matrix and $\lambda_{\max}$ gives the highest eigenvalue. Then consistency index (CI), which measures the inconsistencies of pair-wise comparisons, is calculated as:

$$CI \frac{(\lambda_{\max} - n)}{(n - 1)} \qquad (5.5)$$

Knowing the consistency index, Prof. Saaty proposed to compare it with the appropriate index, and thus, that appropriate consistency index was called random consistency index (RI) and random index (RI) along with numbers of elements (n) used is shown in Table 5.11.

Lastly calculating consistency ratio (CR), which is a comparison between consistency index (CI) and random consistency index (RI). Generally, if CR is less than 0.1 (10%), the judgments are consistent and acceptable. Formulation of CR is:

$$CR = \frac{CI}{RI} \qquad (5.6)$$

where random index (RI) denotes the average RI with the value obtained by different orders of the pair-wise comparison matrices. The values of consistency test are given in Table 5.12.

5.4.7 Priority weights for alternatives

The chances of successfully implementing TI measures in an organization increase only if attributes (sub-objectives) present are strong. Priority weights are used for measuring the preference of the alternative (success or failure) with respect to an attribute. Thus, if the presence of one attribute is strong in the organization, it is more likely to provide success, as compared to the other attribute which is present but weak. For priority weights, the weight evaluation of each alternative is multiplied in the matrix of evaluation rating by vector of attribute weight and summing over the entire attribute. Thus Table 5.13 summarizes the result of evaluating the

Table 5.12 Results of Consistency Test

Maximum Eigenvalue (λ_{max})	C.I.	R.I.	C.R.
7.531	0.0885	1.32	0.0670

Table 5.13 Priority Weights for Attribute Taken

Attribute	Alternative	Success	Failure	Priority Weight
EC	Success	1	6	0.85
	Failure	1/6	1	0.25
TIC	Success	1	3	0.75
	Failure	1/3	1	0.25
OCC	Success	1	1/2	0.34
	Failure	2	1	0.66
GI	Success	1	4	0.80
	Failure	1/4	1	0.20
PP	Success	1	5	0.83
	Failure	1/5	1	0.17
IP	Success	1	1/3	0.25
	Failure	3	1	0.75
SP	Success	1	6	0.85
	Failure	1/6	1	0.15

possible outcome of the implementation of TI, with respect to each of the seven attributes.

For prediction weight of successful TI implementation,

Decision Index of Success = 0.85×0.329 + 0.75×0.221 + 0.34×0.102 + 0.80×0.174 + 0.83×0.056 + 0.25×0.078 + 0.85×0.039 = 0.72 or 72%.

Thus Decision Index of Failure = 1 − 0.72 = 0.28 or 28%.

From the above calculation, it can be said that the success rate of using TI in the organizations is 72% and failure rate is 28%.

5.5 CONCLUDING REMARKS

This section demonstrates the practical implications based on experience and implementation of AHP in various industries. In the present study, this application has been applied for justification of TI in Indian MSMEs, but this approach can be generalized to any industry. The MSME sector has emerged as a highly vibrant and dynamic sector of the Indian economy over the last five decades. Its contribution is highly remarkable in the overall industrial economy of the country. It contributes in providing job opportunities and acts as supplier of goods and services to large organizations. To improve the flow of credit there is a need to provide low cost finance to the MSME sector, which has limited working capital and is dependent exclusively on finance from public sector banks. The cost of credit in the Indian MSME sector is higher than its international peers.

The technology playing a significant role in the trade patterns of advanced industrial countries is widely accepted. Developing countries are assumed to be technological followers, importing developments from developed countries and using them passively. The development of technology by MSMEs is crucial for them to overcome the fast-changing and fiercely competitive global markets. However, only small numbers of small firms in emerging economies are well equipped to develop necessary technology capabilities and the understanding of technology development is still inadequate. Hence, it is obvious that decision making in technology innovation should be considered from a multiple criteria perspective, and these criteria need to be prioritized, and priorities can be changed over time.

Nowadays many researchers are focusing on using various methods for decision-making to measure qualitative criteria such as AHP and fuzzy under uncertainties. These qualitative methods are being used for nearly all the problems and thus receiving more importance for understanding decision-making models under various uncertain conditions for the organizations. AHP as a precise method for selection/decision making is believed to be useful for managers due to its simplicity in use. Yet again, it is proven that AHP works well in making decision for many types of companies that are involved in different types of problems.

Chapter 6

Structural equation modeling

6.1 INTRODUCTION

Structural equation modeling (SEM) is a multivariate statistical technique to estimate constructs and assess hypotheses testing through a confirmatory approach based on empirical data. According to Shah and Goldstein (2006), SEM is more powerful than other multivariate analysis techniques to evaluate the constructs as the method can: (1) allow studying non-quantifiable variables using constructs underlying the indicators; (2) provide adequate accuracy for hypothesis testing and evaluate an unlimited number of hypotheses; (3) examine the interrelationships between constructs; (4) perform simultaneously multiple regression equations analysis; (5) analyze a massive number of variables having different relationships with several complex models; (6) consider the impacts of ill-measured data through measurement errors of indicators; (7) support validity and reliability tests with several fit indices; and (8) perform comparisons between groups with a more holistic model than traditional statistical analysis techniques. The descriptive and exploratory nature of other multivariable analysis techniques makes SEM the most applicable method for model testing.

6.2 VALIDATION OF FUZZY-BASED TI MODEL THROUGH STRUCTURAL EQUATION MODELING (SEM) USING AMOS

The main objective of conducting this study is to confirm previous study, i.e., evaluation of technological innovation initiatives for Indian Micro, Small, and Medium Enterprises (MSMEs) using fuzzy-based model simulation. As the previous study was done by using the most relevant factors affecting TI like product performance and sales performance and were selected from the vast literature review and further those factors had been simulated by the data given by experts in this field using Fuzzy Logic Toolbox of MATLAB which provides the steps for designing fuzzy interface systems for Fuzzy_TI

DOI: 10.1201/9781003272977-6

model, using graphical tools, and a Simulink block for analyzing, designing, and simulating systems based on fuzzy logic.

6.2.1 Instrument used: AMOS 22.0 software

In the present study structural equation modeling has been done by using the AMOS (Analysis of Moment Structures) 22.0 software. Structural equation modeling (SEM) encompasses such diverse statistical techniques such as path analysis, confirmatory factor analysis, causal modeling with latent variables, and even analysis of variance and multiple linear regressions. SEM is a technique that is able to specify, estimate, and evaluate models of linear relationships among a set of observed variables in terms of a generally smaller number of unobserved variables (Shah and Goldstein, 2006).

SEM is a general term that has been used to describe a large number of statistical models used to evaluate the validity of substantive theories with empirical data. It examines the structure of interrelationship through a number of equations and these equations depict the relationship among the dependent and independent variables, named the constructs, which are unobserved or latent variables represented by multiple indicators. Observed variables are also termed as measured, indicator, and manifest, and researchers traditionally use a square or rectangle to designate them graphically. The response to a Likert scale item, ranging from 4 (to a great extent) to 1 (not at all), is an example of an observed variable. Unobserved variables are termed as latent factors and are depicted graphically with circles or ovals.

Figure 6.1 shows a model of three observed predictors predicting one outcome variable. AMOS introduced a way of specifying models in terms of path diagrams. These path diagrams follow a set of standard conventions.

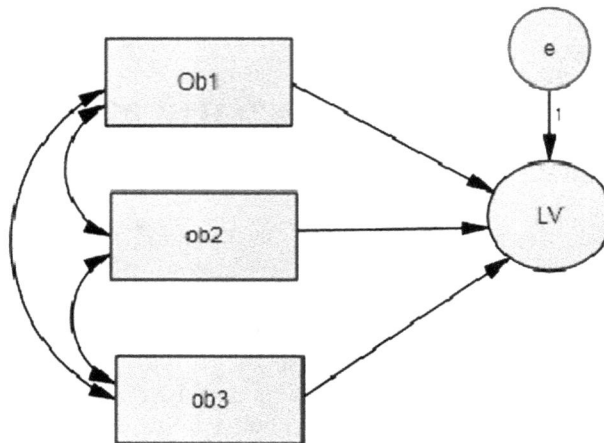

Figure 6.1 Basic model of three observed predictors predicting one outcome variable.

It is an important skill to be able to convert theoretical hypotheses and the data into a path diagram in consisting of:

Rectangle: observed variables (Ob1, Ob2, Ob3), such as items from a questionnaire.
Ellipse: latent variables (LV) that are estimated from observed variables.
Error: error (e), in predicting a variable.
Single-headed arrow: relationships that are predictive.
Double-headed arrow: covariance

The structural equation model consists of two components: the inner model, which shows the linear relationships among the exogenous and endogenous latent variables, and the outer model, which relates each latent variable to its corresponding manifest indicators. SEM has been described as a combination of exploratory factor analysis and multiple regressions (Ullman, 2001). Exploratory factor analysis (EFA) is designed for the situation where links between the observed and latent variables are unknown or uncertain. In contrast to EFA, confirmatory factor analysis (CFA) is appropriately used when the researcher has some knowledge of the underlying latent variable structure. Based on knowledge of the theory, empirical research, or both, the relations are postulated between the observed measures and the underlying factors a priori and then tests this hypothesized structure statistically. It takes a confirmatory rather than an exploratory approach for the data analysis. The present study has been done using the confirmatory factor analysis (CFA) approach using structural equation modeling (SEM).

The core of the SEM analysis should be an examination of the coefficients of hypothesized relationships and should indicate whether the hypothesized model was a good fit to the observed data. In general fit means consistency of two or more factors and it is believed that a good fit among relevant factors will lead to better performance (Shah and Goldstein, 2006). In reference to model fit, researches use numerous goodness-of-fit indicators to assess a model. Some common fit indexes are the normed fit index (NFI), non-normed fit index (NNFI, also known as TLI), incremental fit index (IFI), comparative fit index (CFI), and root mean square error of approximation given by Schreiber et al. (2006) are shown in Table 6.1.

6.2.2 Independent and dependent variables

In reference to the model shown in Figure 6.2 of the fuzzy logic, an extensive literature review was carried out to identify and select the variables for SEM analysis. There are a number of variables that can represent these models, but the variables for the study were carefully selected so that they should bear on the effectiveness of TIIs according to various authors in

Table 6.1 Cutoff Criteria for Several Fit Indexes (Schreiber et al., 2006)

Indexes	Shorthand	General rule for acceptable fit if data are continuous	Categorical data
Absolute/predictive fit Chi-square	χ^2	Ratio of χ^2 to $df \leq 2$ or 3, useful for nested models/model trimming	
Akaike information criterion	AIC	Smaller the better; good for model comparison (non-nested), not a single model	
Browne-Cudeck criterion	BCC	Smaller the better; good for model comparison, not a single model	
Bayes information criterion	BIC	Smaller the better; good for model comparison (non-nested), not a single model	
Consistent AIC	CAIC	Smaller the better; good for model comparison (non-nested), not a single model	
Expected cross-validation index	ECVI	Smaller the better; good for model comparison (non-nested), not a single model	
Comparative fit		Comparison to a baseline (independence) or other model	
Normed fit index	NFI	$\geq .95$ for acceptance	
Incrimental fit index	IFI	$\geq .95$ for acceptance	
Tucker-Lewis index	TLI	$\geq .95$ can be $0 > TLI > 1$ for acceptance	0.96
Comparative fit index	CFI	$\geq .95$ for acceptance	0.95
Relative non-centrality fit index	RNI	$\geq .95$, similar to CFI but can be negative, therefore CFI better choice	
Parsimonious fit			
Parsimony-adjusted NFI	PNFI	Very sensitive to model size	
Parsimony-adjusted CFI	PCFI	Sensitive to model size	
Parsimony-adjusted GFI	PGFI	Closer to one the better, though typically lower that other indexes and sensitive to	
Goodness-of-fit index	GFI	$\geq .95$ not generally recommended	
Adjusted GFI	AGFI	$\geq .95$ performance poor in simulation studies	
Hoelter .05 index		Critical N largest sample size for accepting the model is correct	
Hoelter .01 index		Hoelter suggestion, $N = 200$, better for satisfactory fit	
Root mean square residual	RMR	Smaller, the better; 0 indicates perfect fit	
Standardised RMR	SRMR	$\leq .08$	
Weighted root mean residual	WRMR	$< .09$	$< .09$
Root mean square error of approximation	RMSEA	$< .06$ to .08 with confidence interval	$< .06$

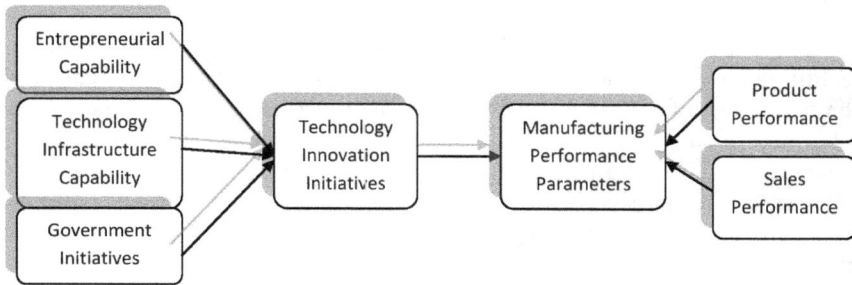

Figure 6.2 Theoretical framework of the SEM_TI model.

the related literature (Lee, 1995; De Toni and Nassimbeni, 1996; Yin and Zuscovitch, 1998; Hayashi, 2002; Schlogl, 2004; Saleh and Ndubisi, 2006; Vohra, 2008; Nanda and Singh, 2009).

Considering these variables the 'TI questionnaire' was designed and sent to selected small scale industrial units for data collection.

Independent variables taken for the study are as follows:

Entrepreneurial capability (EC)
Technology infrastructure capability (TIC)
Government initiatives (GI)

Further, in the this study, the performance parameters or the dependent variables has not been divided into different variables, but only one variable, i.e., 'manufacturing performance parameter' (MPP), has been used for TI model which is mainly used for describing various business performances of the manufacturing organization like product performance and sales performance. This factor has been identified as a significant factor for leading the organization successfully into a turbulent and highly competitive environment.

6.2.3 Structural equation modeling of TI model

The theoretical framework of the TI model is shown in Figure 6.2. The key data required for the model has been obtained using a self-administered questionnaire. The questionnaire used has been divided into two parts. The first part is dedicated to gathering general information about the respondents and their respective organizations such as position, name of the organization, year of inception, annual turnover, etc. Further, the second part of the questionnaire is dedicated to measuring the effectiveness of TIIs in the respective organizations. Each statement in the questionnaire has been designed to extract the respondent's opinion on the above parts in the context of manufacturing performance using a 4-point likert scale.

6.2.4 Screening of the data with preliminary analysis

On the data collected through questionnaire, various data examination techniques were applied like test for skewness and kurtosis, for checking the normality of the data, confirmatory factor analysis (CFA), test for the reliability of the data (Cronbach's alpha), so as to check and increase confidence in the data obtained.

Further, the data was used to construct the SEM_TI model using the structural equation modeling (SEM) using AMOS 22.0 to employ the inter-relationship among the variables used in the study. The result of the SEM analysis allows to understand which variables best explain the constructs and to understand the nature of the relationship between constructs (Mckone et al., 1999).

SEM analysis requires that the data sample should be multivariate normally distributed. Screening of the variables for normality is an important and early step in almost every multivariate test (Tabachnick and Fidell, 2001), as there is no method to test the multivariate normality of the sample data using AMOS or SPSS software. Therefore, in the present study a univariate normality test has been done. Skewness and kurtosis are two important components to measure the univariate normality of the data. As skewness relates to the symmetry of the distribution, a skewed variable is a variable whose mean is not in the center of the distribution (Hatcher, 1994). Table 6.2 (a) and (b) measures the descriptive statistics of all items for independent variable and dependent variable of TI model.

On the other hand, kurtosis measures the spread of data relative to a normal distribution and relates to the peakedness of a distribution (Pallant, 2005). According to Currie et al. (1999), the values of skewness $< \pm 2$ and kurtosis $< \pm 7$ are considered as acceptable. Since the measures of kurtosis and skewness for all items are within the range given by Currie et al. (1999), it is assumed that the distribution of data does not depart from normality.

6.2.5 Confirmatory factor analysis

To measure the determination of the data, i.e., whether it is suitable for confirmatory factor analysis, the strength of the inter-correlations among the items was checked by Bartlett's test of sphericity and using exploratory factor analysis (EFA) the adequacy of the sample size has been checked by Kaiser–Meyer–Olkin (KMO) test (Pallant, 2005). Bartlett's test of sphericity should be significant at $p < 0.05$ for CFA to be considered appropriate, and KMO index should range from 0 to 1, with 0.5 as minimum value for CFA (Tabachnick and Fidell, 2001).

The KMO and Bartlett's tests for the independent and dependent variables are shown in Table 6.3, and the values of the test recommended that the data is suitable to continue with a confirmatory factor analysis procedure.

Table 6.2 (a) Measure of Skewness and Kurtosis for Independent Variables of TI Model

Variable	Items	N Statistic	Minimum Statistic	Maximum Statistic	Mean Statistic	Std. Dev. Statistic	Skewness Statistic	Skewness Std. Error	Kurtosis Statistic	Kurtosis Std. Error
EC	X11	135	1	4	2.99	.885	-.379	.209	-.832	.414
	X12	135	1	4	2.39	.873	-.108	.209	-.757	.414
	X13	135	1	4	3.02	.950	-.681	.209	-.467	.414
	X14	135	1	4	3.08	.864	-.863	.209	.302	.414
	X15	135	1	4	2.03	.897	.382	.209	-.815	.414
	X16	135	1	4	2.29	.771	-.055	.209	-.555	.414
	X17	135	1	4	2.89	1.027	-.613	.209	-.728	.414
	X18	135	1	4	2.30	.947	.283	.209	-.796	.414
	X19	135	1	4	2.93	.883	-.596	.209	-.242	.414
	X110	135	1	4	3.02	.842	-.271	.209	-1.006	.414
	X111	135	1	4	2.15	.797	.176	.209	-.554	.414
	X112	135	1	4	3.06	.853	-.700	.209	-.040	.414
	X113	135	1	4	2.93	.916	-.340	.209	-.885	.414
	X114	135	1	4	2.93	.964	-.475	.209	-.798	.414
TIC	X21	135	1	4	3.09	.851	-.541	.209	-.548	.414
	X22	135	1	4	2.93	.927	-.763	.209	-.110	.414
	X23	135	1	4	2.16	.818	.285	.209	-.439	.414
	X24	135	1	4	2.95	.892	-.346	.209	-.819	.414
	X25	135	1	3	2.18	.732	-.290	.209	-1.082	.414
	X26	135	1	4	2.99	.881	-.370	.209	-.814	.414

(Continued)

Table 6.2 (a) (Continued) Measure of Skewness and Kurtosis for Independent Variables of TI Model

Variable	Items	N Statistic	Minimum Statistic	Maximum Statistic	Mean Statistic	Std. Dev. Statistic	Skewness Statistic	Skewness Std. Error	Kurtosis Statistic	Kurtosis Std. Error
	X27	135	2	4	3.17	.675	-.219	.209	-.803	.414
	X28	135	1	4	2.46	.826	-.070	.209	-.531	.414
	X29	135	1	4	3.12	.820	-.388	.209	-.966	.414
	X210	135	1	4	2.91	.950	-.456	.209	-.751	.414
	X211	135	1	4	2.32	.769	-.019	.209	-.466	.414
	X212	135	1	4	2.94	.929	-.562	.209	-.512	.414
	X213	135	1	4	3.19	.735	-.650	.209	.228	.414
GI	X31	135	1	4	3.15	.718	-.351	.209	-.569	.414
	X32	135	1	4	3.10	.809	-.350	.209	-.929	.414
	X33	135	1	4	2.96	.880	-.447	.209	-.573	.414
	X34	135	1	4	2.30	.764	-.153	.209	-.626	.414
	X35	135	1	4	2.28	.959	.179	.209	-.939	.414
	X36	135	1	4	2.87	.950	-.434	.209	-.739	.414
	X37	135	1	4	2.81	1.052	-.440	.209	-1.005	.414
	X38	135	1	4	2.90	.921	-.504	.209	-.549	.414
	X39	135	1	4	2.85	.927	-.442	.209	-.619	.414

Table 6.2 (b) Measure of Skewness and Kurtosis for Dependent Variable of T1 Model

Variable		N	Minimum	Maximum	Mean	Std. Dev.	Skewness		Kurtosis	
Items		Statistic	Statistic	Statistic	Statistic	Statistic	Statistic	Std. Error	Statistic	Std. Error
T1_MPP	Y1	135	1	4	2.95	.822	-.312	.209	-.598	.414
	Y2	135	1	4	3.03	.914	-.536	.209	-.675	.414
	Y3	135	1	4	3.01	.824	-.664	.209	.121	.414
	Y4	135	1	4	2.94	.879	-.419	.209	-.598	.414
	Y5	135	1	4	2.87	.973	-.484	.209	-.741	.414
	Y6	135	1	4	2.82	.905	-.375	.209	-.613	.414
	Y7	135	1	4	3.05	.892	-.871	.209	.212	.414
	Y8	135	1	4	3.12	.873	-1.054	.209	.739	.414

Figures 6.3–6.6 show the confirmatory factor analysis for all independent and dependent variables of the SEM_TI model. According to Rakowski (1997), items in the variables having the standardized regression weights below than 0.5 should be removed as they will cause the SEM model to unfit. So, it is revealed from CFA done that items X12, X15, X16, X18, and X11 from independent variable EC, items X23, X25, X28, and X211 from independent variable TIC, and items X34, X35, and X37 from independent variable GI should be removed and not to be included in the SEM_TI model.

Table 6.3 KMO and Bartlett's Tests for Independent and Dependent Variables of SEM_TI Model

Variable	Kaiser–Meyer–Olkin Measure	Bartlett's Test of Sphericity	
		Chi-Square Value	P-Value
Entrepreneurial capability (EC)	0.835	1171.870	0.00
Technology infrastructure (TIC)	0.782	731.364	0.00
Government initiatives (GI)	0.782	367.919	0.00
TI manufacturing performance parameters (TI_MPP)	0.855	1115.522	0.00

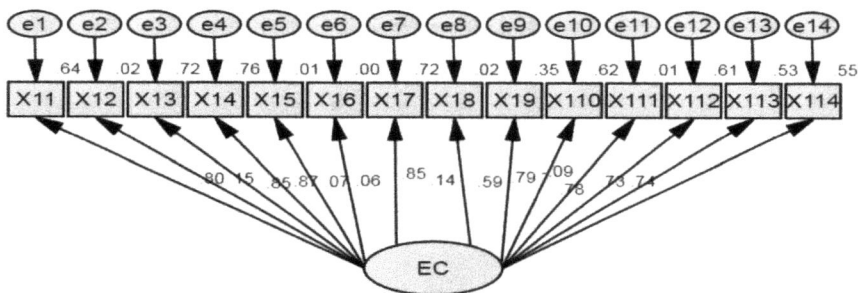

Figure 6.3 Path diagram of the CFA for entrepreneurial capability issues.

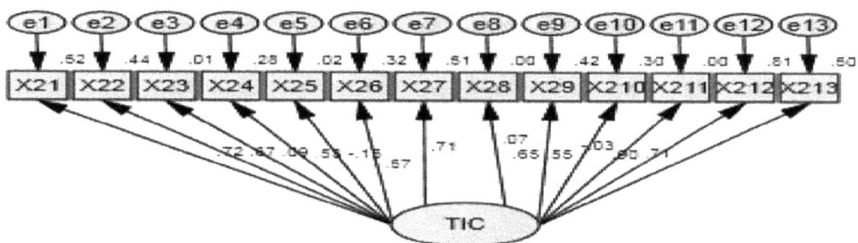

Figure 6.4 Path diagram of the CFA for technology infrastructure capability issues.

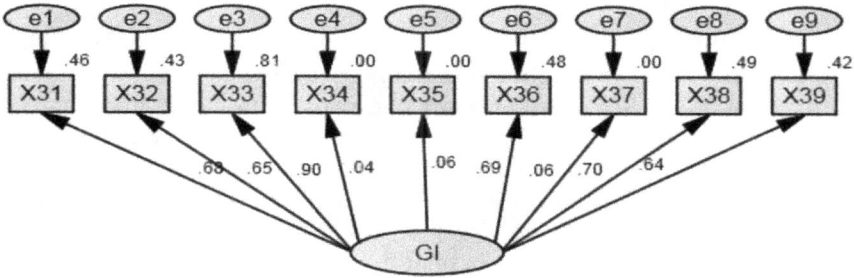

Figure 6.5 Path diagram of the CFA for government initiative issues.

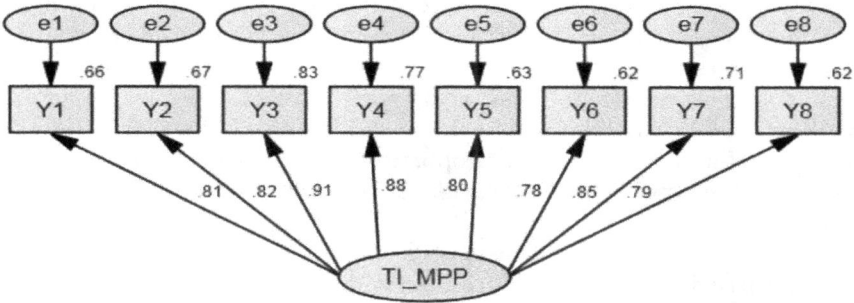

Figure 6.6 Path diagram of the CFA for TI manufacturing performance parameter issues.

Table 6.4 Cronbach's Alpha for Variables of SEM_TI Model

Variables	Items	Cronbach's Alpha (α)
Entrepreneurial capability (EC)	9	0.883
Technology infrastructure (TIC)	9	0.862
Government initiatives (GI)	6	0.720
TI manufacturing performance parameters (TI_MPP)	8	0.754

After removing the items from their respective variables, test for reliability of the data is done using Cronbach's alpha which is a reliability measure of the data for testing internal consistency. The reliability value of 0.7 to 0.8 is an acceptable value for Cronbach's alpha (Nunally, 1978). Table 6.4 shows the values of Cronbach's alpha for all variables which are ranging from 0.72 to 0.86; thus they are considered acceptable and this also increases in confidence of data.

6.2.6 SEM_TI model and result analysis

Figure 6.7 illustrates the full SEM_TI model which is constructed using AMOS 22.0 to build up the relationship between each variable in the study. The SEM_TI model presents the regression coefficients linking the independent construct in the study. AMOS output for the SEM_TI model provides the covariance between independent variables, the ordinary regression coefficient, the error measurement of each independent variable, and the significance level (P-value) for each relationship. The path analysis among all constructs and variables in the model is shown below.

The output of model 1 was obtained and then compared with the cutoff criteria given by Schreiber et al. (2006) for several fit indexes as shown in Table 6.1. It was analyzed that the value of RMR obtained was 0.071. Root mean square residual (RMR) which is the square root of the average squared amount by which the sample variances and covariances differ from their estimates which is preferred as the smaller RMR is the better. The value of goodness-of-fit index (GFI) devised by Jeong and Phillips (2001) was obtained as 0.535. GFI is an alternative to the chi-square test and calculates the proportion of variances that is accounted by estimating population covariance (Tabachnick and Fidell, 2001), which ranges from 0 to 0.95.

The value of adjusted goodness-of-fit index (AGFI) thus obtained 0.463 which is based upon degrees of freedom with more saturated models reducing fit (Tabachnick and Fidell, 2001). When the values of GFI and AGFI come close to 0.95, the model is considered as a perfect fit.

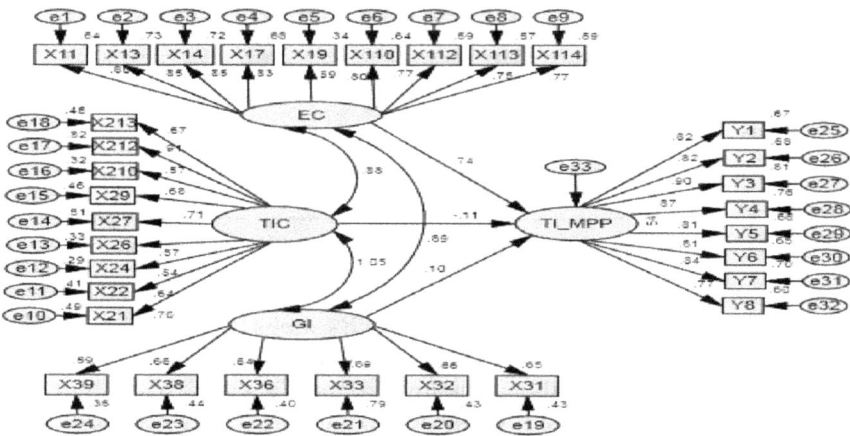

Figure 6.7 Model I: illustrating the full SEM_TI model using AMOS 22.0.

6.2.7 Modification indices of SEM_TI model

Using the modification indices of AMOS 22.0 as shown in Table 6.5, the model was modified accordingly. Modification indices indicate the improvement in fit that will result in the inclusion of a particular relationship in the model. Instead of showing all possible modifications, setting a threshold for modification indices reduces the display of modification indices to a smaller set. In other words, the modification index for a parameter is an estimate of the amount by which the discrepancy function would decrease if the analysis was repeated with the constraints on that parameter removed.

Each time AMOS displays a modification index for a parameter, it also displays an estimate of the amount by which the parameter would change from its current, constrained value if the constraints on it were removed. The modified model and its output are shown in Figure 6.8.

Model fit summary after doing the modification indices and before indices has been collectively shown in Table 6.6. It was observed that after doing the required modifications in the model 1, there has been slight improvement in the model 2 as the value of RMR after doing the modification indices further decreases to 0.064, which is less as compared to RMR value before doing the modification indices. Similarly, the value of GFI increased to 0.584 which is close to 1. The other values have been shown in Table 6.6, which are related to the SEM_TI model as they are required to make the model fit like comparative fit index (CFI), normed fit index (NFI), and relative fit index (RFI).

Table 6.5 Modification Indices for SEM_TI Model

Covariances of items		M.I.	Par Change
e15	<—> e20	84.428	0.286
e5	<—> e7	55.636	0.262
Regression Weights of the Items		**M.I.**	**Par Change**
Y6	<—— GI	11.401	0.345
Y5	<—— GI	4.712	0234
X22	<—— TI_MPP	5.696	0.224
Y6	<—— TIC	7.453	0.224
Y6	<—— EC	6.260	0.176
X21	<—— TI_MPP	4.483	0.171
Y4	<—— TIC	5.024	−0.154
X29	<—— TI_MPP	4.929	−0.176
Y4	<—— GI	5.369	−0.198

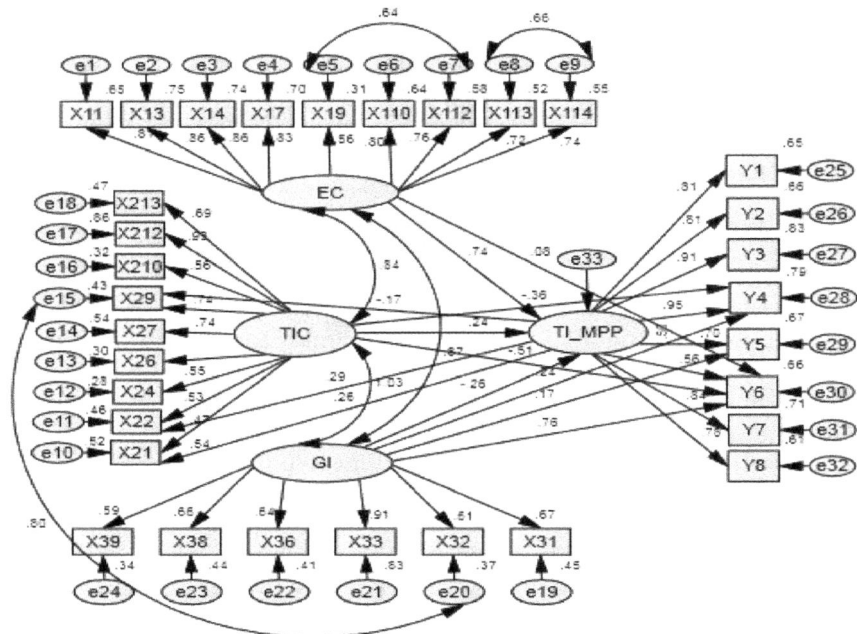

Figure 6.8 Model 2: path diagram of SEM_TI model after doing the modification index.

Table 6.6 SEM_TI Model Statistics

Model Fit Summary	Before Modification Indices	After Modification Indices	Recommended Value for Model Fit
CMIN/df	4.037	3.396	x2/df < 3.0
Degrees of freedom	458	446	Smaller is better
Probability level	0.00	0.00	
Root-mean-square residual index (RMR)	0.071	0.064	Smaller is better; 0 indicates perfect fit
Root-mean-square error of approximation (RMSEA)	0.151	0.134	<0.08
Baseline comparisons			
Goodness-of-fit index (GFI)	0.535	0.584	>0.95
Adjusted goodness-of-fit index (AGFI)	0.463	0.507	>0.95
Comparative fit index (CFI)	0.679	0.753	>0.95
Incremental fit index (IFI)	0.682	0.756	>0.95
Normed fit index (NFI)	0.617	0.686	>0.95
Relative fit index (RFI)	0.585	0.651	>0.95
Tucker–Lewis index (TLI)	0.652	0.726	>0.95

6.3 CONCLUDING REMARKS

Technology innovation is considered to be essential for the survival and growth of individual firms. Technological innovation in the area of both product and process technologies is taking place at a very fast pace. The only way a country can survive is through technological innovation. It is the means to not only increase production, improve quality, and develop new products but also increase competitiveness, expand export, and ultimately ensure uninterrupted economic growth.

Based on the above concept a study was done for modeling Fuzzy_TI model using Fuzzy Logic Toolbox of MATLAB which provides the steps for designing fuzzy inference system using graphical tool, and a Simulimk block for analyzing, designing, and simulating systems based on fuzzy logic. After applying various 'if then' rules which were obtained from the experts of different manufacturing companies, to the inference engine of the fuzzy system, it was analyzed that the Fuzzy_TI model proved to be helpful for the manufacturing organizations for obtaining higher business performance.

To validate this study empirically, the SEM_TI model was formed with the structural equation modeling (SEM) using AMOS 22.0. The data required for forming this model was obtained from respondents of different small scale manufacturing organizations through self-administrated 'TI questionnaire'. Various important factors required for the SEM_TI model, i.e., entrepreneurial capability (EC), technology infrastructure capability (TIC), government initiatives (GI), and manufacturing performance parameters were formulated from the questionnaire and from the extensive literature review. Further various data examination techniques like test for skewness and kurtosis, i.e., to check the normality of the independent and dependent variables data were applied on the SEM_TI model.

Through confirmatory factor analysis (CFA), various items which were affecting for the model to unfit were removed from independent and dependent variables. Lastly using AMOS 22.0 software structural equation modeling (SEM) of technology innovation initiatives was done and their statistics data before and after modification indices were compared. After comparing the SEM_TI model, it was observed that obtained values are near to fit values, which implies that organizations implementing technology innovation initiatives, i.e., entrepreneurial capability (EC), technology infrastructure capability (TIC), government initiatives (GI), are getting better business performances in terms of productivity, market share, profitability, and even cost of product and improved product life cycle of the product. Hence, this also supports our previous work.

Chapter 7

Case studies

7.1 INTRODUCTION

The information collected through industrial survey has been validated through the case studies by using various techniques such as SWOT analysis, SAP (Situation–Actor–Process) analysis, and LAP (Learning–Action–Performance) analysis.

The industrial units considered for empirical analysis fall under four major areas, i.e., cutting tool, machine tool, hand tool, and auto component units. Therefore, the industrial units for conducting case studies were selected on the following criteria:

- The organization is forthcoming and cooperative for getting conducted the case studies.
- The organization has enough tools, machinery, and activities going so as to carry out case study related to manufacturing operations considered in empirical analysis.

The case studies have been performed with respect to the most significant technology innovation initiatives (as per data analysis results) in order to ascertain the exact status of these factors in Micro, Small, and Medium Enterprises (MSMEs).

7.2 INTRODUCTION TO INDUSTRY 'A'

Industry 'A' is a well identified name in manufacturing and supplying an extensive array of thrust bearings in Gurgaon, Haryana (India). It adds in enriching the value of well-engineered mechanical equipment, used and produced in different areas, with their ultimate production of bearings.

7.2.1 Quality policy

Industry 'A' is committed to make magnificent quality bearings for its customers. Since 2001, bearing manufacturing has been our sole business. We

DOI: 10.1201/9781003272977-7

have obtained substantial rise in the industry for quality products. High quality raw materials with high tech machinery are used for manufacturing the stock preserving best qualities of our items since our beginning and keen on enhancing it on a regular basis.

7.2.2 Mission

Our mission is to set benchmarks of attaining optimum clients' satisfaction by meeting their requirements beyond the expectations. For this, we are consistently delivering quality products that are made after conducting extensive studies and understanding the specific requirements of customers.

7.2.3 Product range

Spherical thrust bearing:

- These bearings have spherical raceway in the housing washer and barrel-shaped rollers obliquely arranged around it.

Tapered roller bearing:

- These bearings are designed such that their conical rollers and raceways meet at a common apex on the bearing axis.

Deep groove ball bearings:

- Optimized internal geometry
- Better ball quality

Kingpin bearings:

- They consist of two tapered thrust race, rollers, cage, and an outer retainer which holds components.

The case studies have been performed with respect to the most significant technology innovation initiatives (as per data analysis results) in order to ascertain the exact status of these factors in Indian MSMEs. The information collected through industrial survey has been validated through the case studies by using various techniques such as SWOT analysis, SAP (Situation–Actor–Process) analysis, and LAP (Learning–Action–Performance) analysis. The following section describes the detailed analysis carried out at Industry 'A'.

7.2.4 SWOT analysis at Industry 'A'

Table 7.1 represents the detailed results of SWOT analysis carried out at Industry 'A'.

Table 7.1 SWOT Analysis at Industry 'A'

Aspect	Strengths	Weakness	Opportunities	Threats
Entrepreneurial capability	• Entrepreneur at *Industry* 'A' done certified course in automotive engineering from IMI, Bangalore. • Awareness regarding changing market scenario. • Have sufficient knowledge and experience regarding production activities.	• Lack of management to motivate employees. • Inadequate attention to marketing strategies.	• To plan new projects depending upon market requirement. • To surpass competitors by penetrating into new markets. • To make use of various government schemes for firm growth.	• Growing market requires continuous attention toward modification of routine operations. • Continuous training is required to enhance skill level of workers as well as entrepreneur.
Technological infrastructure capability	• Located near large industries. • Have sufficient machines and equipments for routine operations. • Large customer base. • Wide range of products.	• Unavailability of raw materials at reasonable price. • Difficulty in getting loans. • Lack of finance for staring new projects.	• To become direct supplier for nearby medium and large firms. • To retain skilled manpower. • To make use of latest technology so as to meet changing market demand.	• Large number of firms making similar kind of products nearby. • Increase in investment is required to acquire new technology.
Government initiatives	• Providing subsidized information regarding policies and procedures to help small firms. • Government financial institutions provide credit to small firms for effective functioning of routine operations.	• Lack of awareness among entrepreneur of small firms due to ineffective implementation of plans by Govt. • Inadequate supply of power.	• To make sure that existing policies/ schemes should reach small firms in time. • To make direct link between small and medium as well as large firms.	• Small firms sometimes hesitate to make use of available facilities due to lengthy procedures followed by Govt. Institutions. • Inability of small firms to handle tax structure formed by Govt. due to lack resources and income.

7.2.5 SAP analysis of Industry 'A'

7.2.5.1 Situation

- Industry 'A' is located in IMT Manesar, Gurgaon, where a huge number of large firms are located which enhances its customer base.
- The unit consists of experienced labor and most of the workers have been working with the firm for the last 5 years.
- The testing machines include roundness tester, coordinate measuring equipment, and rockwell hardness tester. Gauges are used to check accuracy of finished products.
- The unit consists of one production manager, four skilled personnel, and seven workers to carry out the production activities smoothly.
- Entrepreneur devotes full time for smooth running of the firm.
- The industry faces the problem of availability of raw materials at reasonable price.
- Working hours in the unit are not fixed as it depends upon the number of orders to be executed. No extra payment is made to the workers for 1–2 hours of overtime.
- The unit is equipped with two diesel-operated electric generators for smooth working of machine and equipment.
- The unit is still required to develop its website for better recognition in the market.
- The unit is offered with tough competition from local counterparts.
- Production capacity of unit has increased during last 1 year.
- The unit has not purchased any latest machinery/equipment for last 3 years.
- Product layout is employed in the unit as same kinds of products are manufactured in large quantity.
- There is lack of training facilities to enhance skill and knowledge base of the employees.

7.2.5.2 Actor

- Entrepreneur of 'TAE', production manager, workers, suppliers, and customers of the industry are working as 'actors'.

7.2.5.3 Process

- Entrepreneur of the unit is involved in almost all the activities carried out during manufacturing of the products.
- The available manpower and machinery are effectively utilized to convert raw material into finished product with 8 hours of working every day.
- Some of the operations are outsourced such as heat treatment.
- The unit employs traditional methods of manufacturing for making required products.

- There is no provision for in-process inspection as only finished products are inspected for their accurate size.
- Continuous production is required to fulfill customer orders on time due to shortage of skilled manpower.
- Regulatory policies like tax structure are affecting performance of the unit in terms of inadequate supply of power and raw material.
- The unit is working toward fulfilling existing customer requirements and no emphasis is given to extend customer base.
- The unit faces difficulty in fulfilling large quantity of orders as less manpower is available for overtime.
- Downtime in the unit is higher due to unavailability of alternatives during machine breakdown.
- The unit is trying to improve the implementation of new plans and strategies in order to enhance productivity.

7.2.6 LAP synthesis of Industry 'A'

- Entrepreneur of the unit should focus on effective utilization of various policies/schemes offered by government for MSMEs.
- There is a need to enhance skill level of workforce through on-the-job training activities.
- Motivation in terms of rewards and incentives is required to attract new and retain existing skilled personnel.
- Entrepreneur of the unit need to make collaboration with other firms in order to expand business activities.
- There is a lot of space for the unit to expand its customer base due to availability of potential customers. Therefore, the unit needs to be well equipped with latest technology infrastructure to meet changing customer demand.
- The unit should implement new types of production processes to enhance productivity.
- Effective documentation is required to implement and standardize the production processes.
- Marketing strategies should be employed to promote sales of the firm.
- The unit is dependent upon selected number of suppliers for raw material purchase. It is required to increase list of suppliers so as to get raw materials at reasonable price and in specified time.
- Government should provide training facilities for small firm's employees as well as entrepreneur at reasonable price.
- Small firms are largely dependent upon banks and other financial institutions to overcome financial constraints.
- Government should provide adequate finance to small firms at low interest rate to enhance productivity and growth.

7.3 INTRODUCTION TO INDUSTRY 'B'

Industry 'B' was established in 2007 as manufacturing, trading, and supplying of tractor components, combine components, and fabricated components to various customers in the northern region and is situated at Mohali, Punjab (India). The manufacturing unit is well armed with all the essential machinery, devices, tools, and equipment in order to manufacture all the components as per requirements of customer with assured quality.

Industry 'B' has gained a good base of clients across the country due to their punctual delivery, positive records, easy mode of payments, and fair business policies. The product category is highly appreciated among the clients due to precise dimensions, easy installation, corrosion-resistance, sturdy design, and durability.

As per the particular needs of the clients, Industry 'B' provides these components in a variety of sizes, shapes, dimensions, materials, and other such specifications to choose from. Quality is the main motto of the unit, so a more emphasized atmosphere is created among the team. Uncompromising quality and exacting specifications are their forte and this is what endows them with the trust of millions.

7.3.1 Milestones

Industry 'B' has created several milestones and a few of these are provided in Table 7.2.

7.3.2 Product range

Machined components:

Titanium (titanium couplings/titanium washers):

- Rust proof
- Dimensional accuracy

Table 7.2 Milestones Established by Industry 'B'

Year	Milestone
2013	Developed machining methods for Titanium
2013	Registered as certified vendor of SML ISUZU
2014	Official website launched
2014	Developed operations for rolling mill parts
2014	Registered as certified vendor of FMGIL
2015	Technology upgraded to NC operation
2016	Registered with Mohali Chambers of Industries and Commerce

- Unmatched quality
- Light weight
- More strength

Polypick (nylon/plastic machined components)

- Easy to install
- Rust proof
- Durable

Fabricated components

Movable stand/trolleys/fixture/components

- Smooth finishing
- Rust proof
- Good appearance
- Smooth performance
- Easy to fit
- Low maintenance
- Cost effective
- Fine finish
- Quality approved

Grinded components

Surface-grinded components

- Rugged construction
- Abrasion resistance
- Durability
- High strength

Fasteners

- Cost effective
- Fine finish
- Quality approved
- Rugged construction
- Precise designs
- Optimum quality

Rolling mill spares

Twist pipe/shearing blades/die coupler

- Easy to install
- High load-bearing ability
- Long service life

- Rust proof
- Cost effective
- Fine finish
- Quality approved

7.3.3 SWOT analysis at Industry 'B'

Table 7.3 presents the detailed results of SWOT analysis carried at Industry 'B'.

7.3.4 SAP analysis of Industry 'B'

7.3.4.1 Situation

- Industry 'B' has variety of product range as more than 23 products are manufactured and machined at the location.
- A small dedicated team consisting of entrepreneur itself, one manager-cum-supervisor, and nine workers including one foreman is focused on manufacturing the products according to customer's requirements.
- Industry is having traditional layout and outlook as setup in the pre-liminary stage.
- Production capacity has been increased by the firm over last few years. In the beginning, only three products were made which are now increased up to 23 products.
- Almost half of the labor is employed as casual workforce and there is often high absenteeism which affects the productivity of the unit.
- Suppliers are connected to industry by means of telephone, fax, and e-mail. Main raw materials used in the industry are gunmetal, titanium, cast iron, and mild steel.
- Cleanliness and other environment-related activities are considered sincerely in the unit.
- Very little support is provided by government agencies regarding awareness of technology up-gradation, procurement of raw material at reasonable price, and hire or purchase of imported goods and machines.
- Fuel and power tariffs are high as compared to neighboring states. Cost of electricity increased from 337p/kwh to 574p/kwh in the time span of 2005 to 2014 whereas fuel charges were raised from INR 37.45 (in 2010) to INR 51.91 (in 2015).
- Lack of finance to start new projects is the major problem for the unit and presently a loan of Rs. 12 lakhs is still under processing for sanction.
- There is very tough competition for Industry 'B' in domestic as well as national markets in terms of new design, quality, variety, and delivery/service-related aspects which create value of the industry in view of customer.

Table 7.3 SWOT Analysis at Industry 'B'

Aspect	Strength	Weakness	Opportunities	Threats
Entrepreneurial capability	• Strong work ethic. • Strong management team. • Tactic knowledge from prior working experience.	• Weak market image. • No clear strategic direction. • Lack of collaboration with government agencies.	• Participation by entrepreneur leads to diversification of market. • Identification of more growth opportunities as well as technology gaps. • Form strategic alliances.	• Entry of strong competitors. • Limited amount of skilled labor to implement new technologies. • Improvement on current infrastructure.
Technological infrastructure capability	• Adequate finance for routine activities • Machinery and equipment are in place and working. • Effective utilization of available resources.	• Information system for marketing and promoting products is not properly utilized. • Softwares for drafting, designing, modeling, and analysis in very limited use. • Lack of local heat treatment facilities	• Target new market. • To introduce new technology and products. • To do business operations with information systems (e-purchasing, use of RFID, bar codes, etc.).	• Changing market demand. • Inability to start new projects. • Obsolete process technology in use.
Government initiatives	• Providing lab facilities for R&D initiatives at subsidized rates. • Providing subsidized information for MSMEs.	• Lack of capital subsidy schemes by government for technology development • Delivery delays caused by slow government procedures. • Lack of cheap and reliable power supply.	• To offer new and improved services. • To implement new policies and measures to support innovation initiatives. • To make aware MSMEs regarding various schemes provided by government	• High interest rates loans provided by financial institutions. • High alloy surcharge on raw material severely increases the input costs. • Increased regulation environment.

- Same labor laws are applicable on all the manufacturing sectors, i.e., micro, small, and medium enterprises irrespective of their size, manpower strength, etc. Following these laws is rather difficult for small scale industry.
- Unit is located in an industrial area where it can avail its basic requirement easily and in time. Sewerage, sanitation, and electric supplies are connected in a planned way.

7.3.4.2 Actor

- Management of Industry 'B', manager, foreman, workforce, and customer's feedback for the unit are the 'Actors'.

7.3.4.3 Process

- Industry 'B' has a capable team and mechanism to convert raw materials into useful products. Every product is accompanied with product drawing for effective utilization of production operations.
- Industry 'B' manages all the records through a computerized management system. There are different types of parts which need to be identified for better tracking and procurement during manufacturing. The record of supplied parts, inventory, and customer orders is maintained easily using a computerized management system.
- Production is limited to conventional manufacturing techniques. Nonconventional machining processes are not yet installed in the unit.
- Industry 'B' adopted cost estimating software in 2011. It comes with manufacturing data and cost models that help the company to be more consistent in their estimates regarding cost of raw materials and finished products.
- Bill of material is also prepared using software which reduces consumption time and increases productivity.
- Overtime is also paid to workers as there is high production demand. However, this activity is not performed on a regular basis but it costs almost Rs. 2000/– to 3500/– per head/month.
- Finished products are transported through trucks/tempos to the nearest road or rail transport carrier for timely delivery to the customers.
- Top management at Industry 'B' is committed toward betterment of its employees as well as suppliers and customers. They seek credibility in leadership, quality, and flexibility in functions at various stages in the unit.
- Two generator sets have been installed at the unit to encounter power supply-related problems.
- Industry 'B' is not able to get benefits provided by government because of lack of awareness regarding the same.

- Industry 'B' is trying to enhance its customer by offering timely and quality products in the market on a continuous basis.

7.3.5 LAP synthesis at Industry 'B'

- Entrepreneur at Industry 'B' needs to be aware of various government schemes so as to get maximum benefits out of them.
- Efforts should be made to replace traditional methods of manufacturing with non-conventional machining processes.
- Although the industry commenced the use of time saving software, the software related to designing for generating process plans, engineering drawings, and layouts would also be beneficial.
- Marketing strategies are required to be effectively implemented in the unit to be able to compete with competitors in the sector, regardless of their size.
- Industry 'B' needs to promote training activities for their employees to enhance their knowledge and skill levels for various machine operations.
- Employees at Industry 'B' are required to be motivated to achieve organizational goals in an effective manner. Reward schemes need to be introduced for motivation and effective utilization of human resources.
- Special consideration should be provided to produce quality products to enhance customer base. Separate inspection departments must be incorporated to ensure in-process quality.
- Strategic financial plans should be employed for effective utilization of funds acquired from financing agencies.
- There is a lack of support and co-operation with research organizations to promote technology-related projects. Industry 'B' should try to collaborate with research organizations to attain technological advancement.
- Government should provide small scale industry with uninterrupted power supply to help increase their production.

7.4 SUGGESTIONS AND RECOMMENDATIONS

The following section presents some suggestions and recommendations for entrepreneurs, financial institutions, and government to resolve various issues related to problems faced by MSMEs.

7.4.1 Suggestions to the entrepreneurs

- The entrepreneurs should take proper training through the government and non-governmental agencies before starting a unit; this enables the entrepreneurs to protect their units from sickness.

- The entrepreneurs should employ latest techniques of production and skilled labor so as to improve the quality of the products and marketing.
- As the competition is found to be a major problem in many units, the entrepreneurs should try to divert to less competitive areas and before they venture, they should analyze the demand.
- Low level of education should not deter one from starting an industrial venture, though; it is a fact that people with higher educational levels are finding their entry into industry easier. Moreover, higher the level of education, the greater is the chance to start a venture as a first generation entrepreneur.
- The spread of schooling has cut across the business of religion. None of the entrepreneurial religions are placed in a disadvantageous position, by comparison.
- Low level of parental education does not prove a hindrance to entrepreneurship.
- Urban background is not a pre-condition of industrial entrepreneurship.
- What is an ambition for one entrepreneur may be a compulsion for another. It is the entrepreneurs' attitudes that ultimately make the difference.
- Previous experience in manufacturing and encouragement of family members/relatives/friends facilitates entrepreneurship.
- Ambitions motivate men. It activates men, broadens their vision, and makes life more meaningful.
- Many of the entrepreneurs expect a lot from the state government and other non-government agencies, but never expect its exact fulfillment.
- Previous experience or employment in the industry should form a basis for selecting the right type of industries.
- For starting a venture, the availability of enough finance is the most important factor. Without it, the idea to start business or venture will always remain a simple wish.
- One should have some basic and essential managerial skills in the functional areas like finance, production, and marketing for entering into industrial entrepreneurship.
- Labor should be given full opportunity of being trained. The problem of absenteeism of labor needs to be looked into with a humane approach. There should be employer–employee friendly relationship inside the industrial unit.
- Entrepreneurs need to re-think about their banking habits. Banks are here to help the entrepreneurs but it does not mean that these helps from the banks are taken for granted. Timely repayment of bank loans is the need of the hour.
- The small scale industrial units should maintain proper books of accounts. Statutory obligation should be imposed on the units

to maintain and prepare their books of accounts by professional accountant.

- Everyone cannot be a successful entrepreneur. An individual must have certain values and traits to be a successful entrepreneur. The traits and values are need for achievement, need for power, positive work value, moderate job anxiety, risk-taking propensity, internal control orientation, high level of aspiration, and preference for participative and nurturing-task styles of leadership.

7.4.2 Suggestions to the government

- The government must provide efficient and effective consultancy services to the entrepreneurs.
- Unhealthy competition among the small units as well as large units should be discouraged as far as marketing problems are concerned. The state government needs to be active in this regard. As a sign of encouragement to local entrepreneurs, government departments should procure products produced by these entrepreneurs.
- Raw material banks need to be opened up in states. Scarcity of raw materials and their high prices as a result of it are the main problems of raw materials.
- Both the central and state governments should give wide publicity so as to reach the information to all the entrepreneurs about policies, incentives, schemes, programs, etc., relating to small scale industry.
- Arrangements may be made by the government to ensure the supply of trained and professional managers for the small scale sector.
- To facilitate the MSME sector to garner resources, it is imperative that a separate trading exchange be set up exclusively for the MSMEs.
- Provide special incentives for encouraging larger flow of venture capital and private equity funds into the sector.
- There is an urgent need to devise measures to tackle the problem of loss of fiscal benefits when the micro and small scale units graduate into larger units, etc.
- Unutilized capacity of an industry is an index of its problems and all the problems faced by industry leads to underutilization of installed capacity. Power scarcity is the main reason for underutilization of capacity. Every possible step should be taken to improve the power condition of the state on a priority basis.
- As far as possible, in order to reduce the competition from the large sector, the small scale industrial units should operate in the areas reserved for them. Similarly, more number of items should be reserved for the exclusive production of the small scale sector.

7.4.3 Suggestions to banks and other financial institutions

- Banks and other financial institutions must undertake a careful project appraisal before assisting an industrial unit. At the same time, financial transactions of their clients must be closely monitored by sending questions and information instead of just receiving periodical but outdated returns from them.
- Banks must supervise the assisted units relating to the proper utilization of funds provided to them in terms of loans.
- Application procedures and approval criteria should be made simple and quick loan approvals should be done at the branch level.
- One or two staff members of the banks and other financial institutions must be always in charge of periodical inspection of the assisted units so as to ensure its efficiency and proper review of production schedules, stock of raw materials, and finished goods.
- The financial agencies must treat loan seekers as customers and not beggars.
- The level of confidence of both entrepreneurs and bankers can be improved by constant follow-up and monitoring. It will help in developing a feeling of partnership among bankers and entrepreneurs in the growth of small enterprises.
- Any sort of discrepancies in collecting information or from the unit should not be tolerated. The service of efficient financial analysts must be available to each bank branch at short notice.
- Timely and adequate finance extending up to the operational cycle of the activity must be available to the entrepreneurs.
- As a general rule, banks and other financial institutions must nominate directors in the industrial units, more particularly those which are likely to detect sickness.
- Utmost care must be taken in financing stocks of raw materials and finished goods which are often affected by sharp price fluctuations.
- The commercial banks and financial agencies may establish more small scale industrial specialized branches at least one in every district headquarters to cater to the financial needs of small entrepreneurs.
- In case of overdrawing, the assisted firms need to be suggested about its underlying problems and should also be ensured that remedial measures are initiated.
- Banks need to be extra careful when they have to provide large funds to neglect and priority sectors. Such lending should not be at the cost of financial prudence.
- Banks need to re-think about their loan-giving policies to the entrepreneurs. Shortage of working capital is the main factor responsible for slow commencement of an industrial unit. So, proper handling of this problem is very important.

7.5 CONCLUDING REMARKS

Case studies have been conducted in four small scale manufacturing organizations which are engaged in the process of adopting technology innovation. The role of technology innovation initiatives in enhancing manufacturing performance has been analyzed in a selected class of industry. The technology innovation initiatives such as entrepreneurial capability, technology infrastructure capability, and government initiatives have been examined to understand and assess their role in achieving manufacturing performance.

The practical difficulties and constraints faced by the small firms due to globalization and dynamic market environments with necessity of technological innovation have also been examined. It has been observed that effective implementation of TIIs effectively contribute toward enhanced manufacturing performance despite numerous problems faced by small firms. Learning issues for each case study have been synthesized and recommendations/suggestions are also made.

Chapter 8

Conclusions and recommendations

8.1 INTRODUCTION

Micro, Small, and Medium Enterprises (MSMEs) are the fountain heads of several innovations in manufacturing and service sectors and by promoting MSMEs into rural areas will assist India to become a developed country. The Indian market is rising rapidly and Indian entrepreneurs are building remarkable progress in various industries such as precision engineering design, manufacturing, food processing, textile and garments, retail, pharmaceutical, IT, agro, and service sectors.

Although the MSME sector has admirable contribution to the Nation's economy, it does not obtain the requisite support from the concerned financial institutions, government departments, banks, and corporates, which act as a handicap in becoming more competitive in the national as well as international markets.

Technological progressions have increased greatly the competition impelled by the globalization of the world economies. It is a remarkable and in certain instances worrying situation because MSMEs play an important role in most economies, including in India, in which they comprise the largest business block and offer the bulk of employment. There have been several studies which highlighted certain factors contributing toward the technological innovation initiatives of the firms, especially MSMEs. Further, only a few empirical studies have been found to support the theoretical findings. There are remote cases where the relative impact of technological innovation initiatives on performance enhancements, especially in the context of the Indian small scale manufacturing sector, has been reported.

With this backdrop, the present study has made an effort to focus on empowering small firms. As argued earlier, MSMEs are important contributors to economic development, and it is therefore pertinent to identify technological innovation initiatives to enhance manufacturing performance of the selected class of industry.

DOI: 10.1201/9781003272977-8

8.2 SUMMARY OF THE STUDY

This study is aimed at identifying various critical success factors for overcoming the obstacles to a successful implementation of the technology innovation program in the Indian small scale industry. Moreover, the study illustrates how the synergistic relationship of TIIs can be helpful for the Indian small scale industry to have enhanced manufacturing performance, through a specially designed TI questionnaire.

The major objective of this research is to examine how effective is the support of technology innovation in enhancing manufacturing performance of small firms. Finally, the research culminates with development of a strategic TI implementation model for the Indian small scale manufacturing industry for sustained growth and competitiveness.

8.3 RESEARCH CONTRIBUTIONS

8.3.1 Percent points scored results

The research provides an assessment of prevailing technology-related issues of the Indian small scale manufacturing industry like entrepreneur's capability, technology infrastructure capability, organizational culture and climate, and government initiatives.

(a) The analysis of various issues related to entrepreneurial capability (EC) shows that most of the organizations have entrepreneurs with a good education level (PPS = 75.19), and they have adequate knowledge regarding various government schemes for MSMEs (PPS = 71.11). They have the ability of strategic decision making in identifying the right kind of business and market (PPS = 73.15) as well as the ability to make effective decisions pertaining to business activities (PPS = 68.33). They also strongly emphasize on R&D, technological leadership, and innovative products.

(b) A close analysis of various issues related to technology infrastructure (TI) reveals that most of the organizations get appropriate raw material at reasonable prices (PPS = 78.33) and have technical knowledge and infrastructure to do business operations with information systems (e-purchasing, use of Radio Frequency Identification and bar codes, etc.) as shown by the data (PPS = 75.19). Most of the organizations also have sufficient credit for meeting requirements of routine operations (PPS = 74.26).

(c) The analysis of various issues related to organization culture and climate (OCC) shows that only about half of the organizations have skilled man power to increase competitiveness and suitable growths (PPS = 51.67), and at the same time they provide training to transfer knowledge and skills that are of requisite quality (PPS = 51.48) to a small extent. The extent of use of market and customer feedback into the innovation process is reasonable (PPS = 52.96) and most of the organizations do not have good strength of R&D personnel (PPS = 45.93). The data shows that a reasonable number of organizations focus on employee empowerment (PPS = 65.56) and technological innovation initiatives are also supported by management (PPS = 67.41).

(d) The close analysis of various issues related to government initiatives reveals that the government provides lab facilities for MSMEs to encourage them to speed up technological and new product development projects (PPS = 73.15). Government helps MSMEs in acquiring the latest technology, quality certification, and marketing assistance (PPS = 56.67) and in locating funds for R&D initiatives (PPS = 56.85) up to some extent. These firms are also provided with free or subsidized information regarding the latest trends and technologies in relation to government regulations (PPS = 67.96).

(e) The analysis of various issues related to performance parameters shows that market share of a large number of organizations has increased because of new products (PPS= 75.93), and there is improvement in product life cycle (PPS = 75.19) as a result of TI initiatives. There is considerable improvement in sales due to new products as a percentage of total sales (PPS = 74.63). The TI initiatives have led to the increase in production of new products as a percentage of total products over the last 3–5 years (PPS= 67.78). The analysis of survey reveals that the implementation of TI initiatives improved the technical characteristics and features of existing product range (PPS= 67.59) as well as contributing to the implementation of new processes.

8.3.2 Statistical analysis results

The research has highlighted the contribution of technology innovation initiatives (TIIs) in Indian MSMEs for enhanced manufacturing performance. The empirical analysis of survey reveals that TIIs have yielded considerably significant improvement in Indian MSMEs in terms of improved life cycle of products, reduction in cost of production, mean sales profitability, and increase in market shares.

(a) It is evident from the survey that adoption of TIIs contributed to the sustained competitive advantage of several Indian MSMEs. The successful implementation of TIIs can lead toward realization of strategic manufacturing performance improvements for competing in the highly dynamic global marketplace. These interrelationships can be used to understand the effect of various TII success factors toward realization of organizational objectives of growth and sustainability.

(b) The detailed interrelationship between various TIIs and manufacturing performance parameters indicate that entrepreneurial capability (I1) issues are considerably associated with the manufacturing performance parameters (O1 – product performance (0.33*), O2 – innovation performance (0.20*), and O3 – sales performance (0.25*)). Also, technology infrastructure (I2) issues are found to be closely related to O1 – product performance (0.43*) and O3 – sales performance (0.30*). Government initiatives (I2) issues are correlated with O1 – product performance (0.31*) and O3 – sales performance (0.44*).

(c) The results of T-test (Table 4.21) reveal that various TIIs and manufacturing performance parameters (MPPs) are closely associated since the significant factor 'p' works out to be less than 0.05 in most of the cases. Moreover, the t (critical) value for confidence limits corresponding to n–2 (=133) degrees of freedom and significance level of 5 percent, from statistical t tables, works out to be 1.98, which is lower than the t values obtained for most of the input–output combinations as revealed in Table 4.21. This further validates the high correlation between various TIIs and manufacturing performance parameters.

(d) The results through multiple regression analysis are presented in Table 4.22. The result implies that O1 – product performance is significantly affected by I1 – entrepreneurial capability (EC), I2 – technology infrastructure capability (TIC), and I4 – government initiatives (GI) issues. O2 – innovation performance is significantly affected by I1–EC whereas O3 – sales performance is affected by all TIIs, i.e. I1, I2, I3, and I4.

(e) The results of canonical correlation analysis between TIIs and MPPs are shown in Table 4.23 and indicate strong and significant canonical correlation function (r = 0.750 at F statistics probability of 0.00) between the predictor set of TI implementation dimensions and the criterion set of MPP. The multivariate test statistics have been observed to be statistically significant ($p < 0.001$). The redundancy indices were 0.283 and 0.248 for the dependent and independent canonical variates, respectively. The redundancy index indicates the amount of variance in a canonical variate explained by the other canonical variate in the canonical function. The canonical loadings for the predictor set of various TIIs (I1, I2, I3, and I4) on the dependent variate range from 0.396 to 0.701. The criterion set of manufacturing performance parameter variates (O1, O2, and O3) have also been found to be loaded up to 0.690 on the dependent variate.

8.3.3 Qualitative analysis results

(a) The research has been extended to prove the synergistic suitability of TIIs using fuzzy-based model simulation. For the study, the most relevant factors affecting these drives like product performance and sales performance have been considered as the most important issues to be cared for and data given by experts in simulation uses the fuzzy logic tool box of MATLAB. It also provides the steps for designing the fuzzy interface system and the Simulink block for analyzing, designing, and simulating the system based on fuzzy logic. A continuum of fuzzy solutions for the TI implementation equation is presented using the rule viewer of the fuzzy tool box of MATLAB. The rule viewer displays a roadmap of the whole fuzzy inference process and it is based on the fuzzy inference diagram. The rule viewer allows interpreting the entire fuzzy inference process at once. It also shows how the shape of certain membership functions influences the overall result as it plots every part of every rule. Each rule is a row of plots, and each column is a variable. The rule numbers are displayed on the left of each row. By clicking on a rule number, the rule in the status line can be viewed. The two inputs can be set within the upper and lower specification limits and the output response is calculated as a score that can be translated into linguistic terms. In this instance the order output of 8.52 indicates an 'acceptable system' linguistically from Table 6.8. The rule viewer shows in detail one calculation at a time and in this sense, it presents a sort of micro view of the fuzzy inference system. The results show that if TI measures are used efficiently, their synergistic effect can improve manufacturing performance of an organization.

(b) For validating the Fuzzy study empirically, the SEM has been done. This study uses the confirmatory factor analysis approach using SEM in AMOs 22.0 software to imply the inter-relationship among TII and MPP variables in the study. Model Fit summary of SEM_TI model after doing the modification indices represents that the value of root mean square residual (RMR) decreases to 0.064, which is less as compared to RMR value before doing the modification indices as depicted in Table 6.10. Similarly, the value of GFI increased to 0.584 which is close to 1. This indicates that the model after modifications does provide a better fit w.r.t. normed fit index (NFI) which is equal to 0.686. Thus, the SEM study confirms and validates the TI Fuzzy model, which justifies the previous study.

(c) Further, the justification of TI implementation in Indian MSMEs has been made using the analytic hierarchy process (AHP) by calculating consistency ratio (CR), which is a comparison between the consistency index (CI) and the random index (RI). For the study, the value of CR (0.0670) is coming less than 0.1 (10%) as shown in Table 6.18, which means the judgment considered for the study is consistent and acceptable. Further, Table 6.19 represents that the success rate of using TI in an organization is 72%, whereas the failure rate is 28%.

(d) Finally, a conceptual framework representing the main elements of the technology innovation implementation program for the small scale industrial sector has been developed. The framework has been developed based on relevant literature, sample-based survey, case studies, and qualitative modeling presented in this study. The four independent variables known as TIIs, entrepreneurial capability, technology infrastructure capability, organizational culture and climate, and government initiatives, are aligned to the three dependent variables known as manufacturing performance parameters (MPPs): product performance, innovation performance, and sales performance as represented in the framework. These MPPs collectively lead to enhanced manufacturing performance.

8.3.4 Technology innovation implementation model

The structure of the research model has been developed based on relevant literature, sample-based survey, case studies, and qualitative modeling presented in this study. The four independent variables known as TIIs, entrepreneurial capability, technology infrastructure capability, organizational culture and climate, and government initiatives, are aligned to the three dependent variables known as manufacturing performance parameters (MPPs): product performance, innovation performance, and sales performance as represented in the framework. Furthermore, these MPPs collectively lead to enhanced manufacturing performance. Figure 8.1 shows the research model and possible linkages between different variables.

8.4 MAJOR FINDINGS OF THE STUDY

It can be concluded that in a highly competitive scenario, implementation of technology innovation programs has proven to be the most significant initiative that leads organizations to scale new levels of achievement.

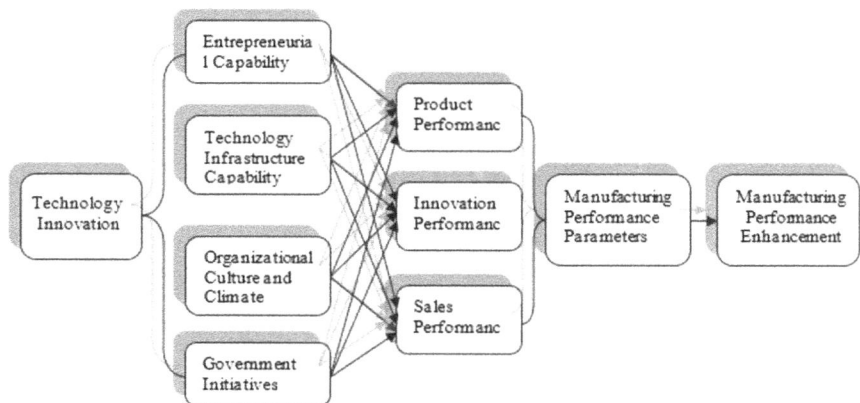

Figure 8.1 A conceptual model of the research.

Accordingly, the following major findings have been drawn out from the elaborative study:

1. Entrepreneurial capability, technology infrastructure capability, and government initiatives have emerged as significant contributors toward enhanced manufacturing performance in the organizations.
2. Entrepreneurial capability is strongly associated with all the manufacturing performance parameters, i.e. product performance, innovation performance, and sales performance, and demonstrated as the most successful factor for implementing technology innovation.
3. Technology infrastructure capability has proven to be an important contributor in improving product performance and sales performance, therefore enhancing the manufacturing performance of the organizations.
4. Government initiatives are significantly linked with product performance and sales performance in the selected class of industry as government assistance for MSMEs provides potential benefits in creating employment opportunities and improving innovation and competitiveness.
5. Organizational culture and climate is closely associated with sales performance in the organizations, which confirms that culture directly affects an employee's ability to work effectively and efficiently and therefore determines the productivity of the firm.
6. The technology innovation-strategy model developed shall prove to be a ready to use tool for enhancing manufacturing performance of Indian MSMEs in a fierce competitive environment.

8.5 LIMITATIONS OF THE STUDY

The research limitations of the present work provide suggestions for future studies. This study has a number of limitations in generalizing the findings across the MSME sector in Indian economy, which include the following:

- The present research is limited to only the cutting tool, machine tool, hand tool, and auto components industry of North India. Factors may vary according to the products of the manufacturing industry like material handling equipments, farm and agri-machinery, bicycle industry, and two wheeler and car parts manufacturing units.
- The present research is carried out in small scale manufacturing units situated in North India only; significance of issues could differ in other regions of the economy.
- In this research, only four significant technology innovation factors (entrepreneurial capability, technology infrastructure capability, organizational culture and climate, and government initiatives) are

studied, whereas other factors (collaborative networks, workforce management, resource management, and external knowledge inflows) might also affect a firm's performance.

- The research suggests a generalized implementation of the TI approach for the Indian small scale industry as a whole and all manufacturing organizations in the study have been treated alike, irrespective of the specific requirements of various small firms. Thus, minor changes have to be incorporated to effectively adopt this program.
- As such no mathematical models or quantitative relationship has been derived to calculate the contribution of various factors in achieving enhanced manufacturing performance.

8.6 SUGGESTIONS FOR FUTURE RESEARCH

While carrying out the study and trying to list its scope, a number of areas have come under focus, where detailed research can be taken up. Such areas demanding attention, further exploration, and analysis through research work are mentioned below.

- The study can be extended to examine the role of other prominent factors like collaborative networks, workforce management, resource management, and external knowledge inflows on manufacturing performance of organizations.
- The scope of the study is limited to the cutting tool, machine tool, hand tool, and auto components industry. Further research can be carried out by considering the different kinds of products manufactured by various industrial units such as material handling equipments, farm and agriculture-machinery, bicycle industry, and two wheeler and car parts manufacturing units.
- All manufacturing organizations have been treated alike, irrespective of the specific requirements of various small firms. Minor changes might have to be incorporated to effectively manage technology innovation in varying situations. Thus, sector-wise analysis can also be conducted to appropriately deal with the varying requirements of different sectors.
- The study is aimed at implementing strategic TI methodology in Indian small scale manufacturing units situated in the Northern region only. Another direction for future research is developing these programs for small firms in other regions of the country.
- A qualitative technology innovation-strategy model has been developed. In future research, a mathematical model for technology innovation strategy could be developed.

References

Abdelsamad, M. H. and Kindling, A. T. (1978), "Why small businesses fail", *SAM Advanced Management Journal*, Vol. 43, No. 4, pp. 24–37.

Abereijo, I. O., Adegbite, S. A., Ilori, M. O., Adeniyi, A. A. and Aderemi, H. A. (2009), "Technological innovation sources and institutional supports for manufacturing small and medium enterprises in Nigeria", *Journal of Technology Management and Innovation*, Vol. 4, No. 2, pp. 82–89.

Acikgoz, A. and Gunsel, A. (2011), "The effects of organizational climate on team innovativeness", *Procedia Social and Behavioral Sciences*, Vol. 24, No. 1, pp. 920–927.

Adenikinju, A. F. (2003), "Electric infrastructure failures in Nigeria: a survey-based analysis of the costs and adjustment responses", *Energy Policy*, Vol. 31, No. 14, pp. 1519–1530.

Adner, R., Helfat, C. E. (2003), "Corporate effects and dynamic managerial capabilities", *Strategic Management Journal*, Vol. 24, No. 10, pp. 1011–1025.

Adukia, R. S. (2012), "An overview of the micro, small and medium enterprises (MSMES) sector", http://www.caaa.in/Image/41MSME.pdf, pp. 1–38.

Afuah, A. (2003), *Innovation Management: Strategies, Implementation, and Profits*, New York: Oxford University Press.

Ahmadi, M. and Helms, M. M. (1997), "Small firms, big opportunities: the potential of careers for business graduates in SMEs", *Education and Training*, Vol. 39, No. 2, pp. 52–57.

Al Ghamdi, S. (2005), "The use of strategic planning tools and techniques in Saudi Arabia: an empirical study", *International Journal of Management*, Vol. 22, No. 3, pp. 376–395.

Aldag, J. R. and Stearns, M. T. (1987), *Management*, Cincinnati: South – Western Publishing Co., Vol. 1, No. 4, pp. 803–812.

Aldehayyat, J. S., Twaissi, N. and Jordan, M. (2011), "Strategic planning and corporate performance relationship in small business firms: evidence from a Middle East country context", *International Journal of Business and Management*, Vol. 6, No. 8, pp. 255–263.

Aldrich, H. E. and Fiol, C. M. (1994), "Fools rush in? The institutional context of industry creation", *Academy of Management Review*, Vol. 19, No. 4, pp. 645–670.

Alkali, M. (2012), "An empirical study of entrepreneurs educational level and the performance of small business manufacturing enterprises in Bauchi State, Nigeria", *Interdisciplinary Journal of Contemporary Research in Business*, Vol. 4, No. 6, pp. 914–923.

Alzaga A. and Martin J. (2006), "A design process model to support concurrent project development in networks of SMEs", *Infrastructures for Virtual Enterprises*, Vol. 16, No. 7, pp. 55–75.

Analoui, F. and Karami, A. (2002), "CEOS and development of meaningful mission statement", *Corporate Governance*, Vol. 2, No. 3, pp. 13–30.

Anderson, W. (2011), "Internationalization opportunities and challenges for small and medium-sized enterprises from developing countries", *Journal of African Business*, Vol. 12, No. 2, pp. 198–217.

Archarya, M. (2008), "The employee customer profit chain", *Harvard Business Review*, Vol. 40, pp. 83–97.

Arthur, J. B. (1994), "Effects of human resource systems on manufacturing performance and turnover", *Academy of Management Journal*, Vol. 37, No. 3, pp. 670–687.

Arthur, M. B. and Hendry, C. (1990), "Human resource management and the emergent strategy of small to medium sized business units", *International Journal of Human Resource Management*, Vol. 1, No. 3, pp. 233–250.

Asian Development Bank. (2009), "Enterprises in Asia: fostering dynamism in SMEs", *Key Indicators for Asia and the Pacific 2009*, http://www.adb.org/sites/default/files/pub/2009/Key-Indicators-2009.pdf, pp. 44.

Asian Productivity Organization. (2007), "Entrepreneurship development for competitive small and medium enterprises", *Report of the APO Survey on Entrepreneur Development for Competitive SMEs(05-RPGE-SUV-41-B)*, http://www.apo-tokyo.org/00e-books/IS-26_SMEs/IS-26_SMEs.pdf

Asiedu, E. (2004), "Policy reform and foreign direct investment in Africa: Absolute progress but relative decline", *Development Policy Review*, Vol. 22, No. 1, pp. 41–48.

Atkinson, J. and Meager, N. (1994), "Running to stand still", in J. Atkinson and D. Storey (eds.), *Employment, the SmallFirm, and the Labour Market*, London: Routledge, Vol. 18, pp. 105–120.

Audretsch, D. B. and A. R. Thurik, (2000), "Capitalism and democracy in the 21st century: from the managed to the entrepreneurial economy", *Journal of Evolutionary Economics*, Vol. 10, No. 1, pp. 17–34.

Bacon, N., Ackers, P., Storey, J. and Coates, D. (1996), "It's a small world: managing human resources in small businesses", *The International Journal of Human Resource Management*, London: Sage Publications Ltd., Vol. 7, No.1, pp. 251–268.

Bailleti A. J., Callahan J. R., DiPietro P. (1994), "A coordination structure approach to the management of projects", *IEEE Transactions on Engineering Management*, Vol. 41, No. 4, pp. 394–403.

Baker, T. and Aldrich, H. (1994), "Friends and strangers: early hiring practices and idiosyncratic jobs", Paper presented at the Fourteenth Annual Entrepreneurship Research Conference, INSEAD, Fountainbleau, France, pp. 65–85.

Bakunda, G. (2003), "Explaining firm internationalization in Africa using the competence approach", *Journal of African Business*, Vol. 4, No. 1, pp. 57–85.

Bala Subrahmanya, M. H. (2007), "Development strategies for Indian SMEs: promoting linkages with global transnational corporations", *Management Research News*, Vol. 30, No. 10, pp. 762–774.

Banks, M. C., Bures, A. L. and Champion, D. L. (1987), "Decision making factors in small business: training and development", *Journal of Small Business Management*, Vol. 25, No. 1, pp. 19–25.

Baporikar, N. and Deshpande, M. V. (2015), "Approaches and strategies of Pune auto component SMEs for excellence", *Journal of Science and Technology Policy Management*, Vol. 6, No. 2, pp. 114–126.

Baral, S. K. (2013), "An empirical study on changing face of MSME towards emerging economies in India", *Journal of Radix International Educational and Research Consortium*, Vol. 2, No. 1, pp. 1–21.

Barber, A. E., Wesson, M. J., Roberson, Q. M. and Taylor, M. S. (1999), "A tale of two job markets: organizational size and its effects on hiring practices and job search behaviour", *Personnel Psychology*, Vol. 52, No. 4, pp. 841–867.

Barro, R. and Sala-I-Martin, X. (1995), *Economic Growth*, New York: McGraw-Hill.

Bartel, A. P. (2004), "Human resource management and organisational performance: evidence from retail banking", *Industrial and Labour Relations Review*, Vol. 57, No. 2, pp. 181–195.

Beaudry, C. and Swann, P. (2001), "Growth in industrial clusters: a bird's eye view of the United Kingdom", *Stanford Institute for Economic Policy Research*, Discussion Paper No. 38.

Becheikh, N., Landry, R. and Amara, N. (2006), "Lessons from innovation empirical studies in the manufacturing sector: a systematic review of the literature from 1993–2003", *Technovation*, Vol. 5, No. 6, pp. 644–664.

Beck T. and Kunt A. D. (2006), "Small and medium-size enterprises: access to finance as a growth constraint", *Journal of Banking & Finance*, Vol. 30, No. 11, pp. 2931–2943.

Beer, M., Spector, B., Lawrence, P. R., Mills, D. Q. and Walton, R. E. (1984), *Managing Human Assets: The Groundbreaking Harvard Business School Program*, New York: The Free Press, Macmillan, Vol. 13, pp. 165–177.

Bennett, R. (1993), "Small business survival", *Natwest Business Handbooks*, London: Pitman Publishing, Vol. 25, pp. 75–90.

Berg, L. D. and Harral M. W. (1998), "The small company route to ISO 9000. Quality Digest", www.qualitydigest.com/july98/html/isosmall.html.

Berger N. and Udell G. F. (2006), "A more complete conceptual framework for SME finance", *Journal of Banking & Finance*, Vol. 30, No. 11, pp. 145–198.

Berry, M. (1998), "Strategic planning in small high tech companies", *Long Range Planning*, Vol. 31, No. 3, pp. 455–466.

Bessant, J., Lamming, R., Noke, H., & Phillips, W. (2005), Managing innovation beyond the steady state. *Technovation*, Vol. 25, No. 12, 1366–1376.

Bhaskaran, S. (2006), 'Incremental innovation and business performance: small and medium-size food enterprises in a concentrated industry environment', *Journal of Small Business Management*, Vol. 44, No. 1, pp. 64–80.

Bi, K. X., Sun, D. H., Zheng, R. F. and Li, B. Z. (2006), 'The construction of synergetic development system of product innovation and process innovation in manufacturing enterprises', in Proceedings of the 13th International

Conference on Management Science and Engineering (ICMSE), Lille, France, 5–7 October, pp. 628–636, ISBN: 7-5603-2355-3.

Biryabarema, E. (1998), "Small scale businesses and commercial banks in Uganda", *Business & Economics*, Kampala Uganda: Makerere University Press.

Bommer, M. and Jalajas, D. S. (2004), 'Innovation sources of large and small technology-based firms', *IEEE Transactions on Engineering Management*, Vol. 51, No. 1, pp. 13–18.

Bosworth, D., (1989), "Barriers to growth: the labour market", in J. Barber, J. S. Metcalfe and M. Porteous (eds.), *Barriersto Growth in Small Firms*, London: Routledge, Vol. 9, pp. 105–120.

Brannen, W. H. (1983), *Marketing and Small Business/Entrepreneurship"*, Washington, DC: International Conference for Small Business, pp. 2–11.

Brautigam, D. (1994), "African industrialization in comparative perspective: the question of scale", in Berman and Leys (eds.), *African Capitalists in African Development*, London: Lynne Rienner Publishers.

Bruque, S. and Moyano, J. (2007), "Organizational determinants of information technology adoption and implementation in SMEs: the case of family and cooperative firms", *Technovation*, Vol. 27, No. 5, pp. 241–253.

Burgelman, R., Maidique, M. A. and Weelwright, S. C. (2004), *Strategic Management of Technology and Innovation*, New York: McGraw.

Buss, D. D. (1996), "Help wanted desperately", *Nation's Business*, Vol. 84, No. 4, pp. 16–19.

Bwisa, H. and Gacuhi, A. R. (1997), *Diffusion and adaptation of Technology from Research Institutes and Universities in Kenya, An Empirical Investigation*, University of Nairobi: Department of Commerce, pp. 87–84.

Cainelli, G., Evangelista, R. and Savona, M. (2004), "The impact of innovation on economic performance in services", *The Service Industries Journal*, Vol. 24, No. 1, pp. 116–130.

Cambridge Small Business Research Centre. (1999), *Survey of Small and Medium Sized Firms, Fourth Report*, Cambridge, UK: University of Cambridge.

Campbell, D., Shrives, P. and Bohmbach-Saager, H. (2001), "Voluntary disclosure of missions statements in corporate annual reports: signaling What and to Whom?", *Business and Society Review*, Vol. 106, No. 1, pp. 65–87.

Cannon, T. (1991), *Butterworth Heinemann: Development and Growth*, Enterprise: Creation, Vol. 75, pp. 225–240.

Cant, M. and Brink, A. (2003), "Problems experienced by small businesses in South Africa", *in 16th Annual Conference of Small Enterprise Association of Australia and New Zealand*, Ballarat (Australia): University of Ballarat, pp. 1–20.

Capitalist Process, London, NY: Mc, Bala Subrahmanya, M. H., Mathirajan, M., Balachandra, P. and Srinivasan, M. N. (2001), *R&D in Small Scale Industries in Karnataka, Research Project Report*, New Delhi: Government of India, Department of Science and Technology.

Cardon, M. S. (2003), "Contingent labor as an enabler of entrepreneurial growth", *Human Resource Management Journal*, Vol. 42, No. 4, pp. 357–373.

Carey, N., Randewich, N. and Krolicki, K. (2011), "Disaster shows flaws in just-in-time production", Retrieved March 22, 2011, from http://www.reuters.com/article/2011/03/21/us-japan-supplychain-sp.

Carmeli, A., Tishler, A. (2004), "The relationships between intangible organizational elements and organizational performance", *Strategic Management Journal*, Vol. 25, No. 13, pp. 1257–1278.

Cassiman, B., Golovko, E. and Martinez-Roz, E. (2010), "Innovation, exports and productivity", *International Journal of Industrial Organization*, Vol. 28 No. 4, pp. 372–376.

Cederfeldt, M. and Elegh, F. (2005), "Design automation in SMEs-current state, potential, need and requirements", in International Conference on Engineering Design, Melbourne.

Chaminade, C. and Vang, J. (2006), "Innovation policies for Asian SMEs: an innovation system perspective", in H. Yeung (ed.), *Handbook of Research on Asian Studies*, Cheltenham: Edward Elgar.

Chandraiah, M. and Vani, R. (2013), "The impact of globalization on micro, small and medium enterprises (MSMEs) with special reference to India", *Innovative Journal of Business and Management*, Vol. 12, pp. 109–111.

Chaston, I. (2012), "Small firm performance: assessing the interaction between entrepreneurial style and organizational structure", *European Journal of Marketing*, Vol. 2, No. 1, pp. 28–35.

Chen, M. and Hambrick, D. C. (1995), "Speed, stealth, and selective attack: how small firms differ from large firms in competitive behavior", *Academy of Management Journal*, Vol. 38, No. 2, pp. 453–482.

Cheng, C. H., Hsiao, L. Y. C. and Tasi, C. J. (2002), "High-tech industry in Taiwan: support for high-tech ventures", *Asia Pacific Tech Monitor*, Vol. 1, No. 1, pp. 1–35.

Chesbrough, H. (2003), *Open Innovation. The New imperative for Creating and Profiting from New Technology*, Cambridge, MA: Harvard Business School Press, pp. 56–61.

Chesbrough, H. (2006), Open innovation: a new paradigm for understanding industrial innovation. In H. Chesbrough, W. Vanhaverbeke, and J. West (Eds.), *Open Innovation: Researching a New Paradigm*. Oxford: Oxford University Press, pp. 1–12.

Chesbrough, H. and Bogers, M. (2014), "Explicating open innovation: clarifying an emerging paradigm for understanding innovation", in H. Chesbrough, W. Vanhaverbeke, & J. West (eds.), *New Frontiers in Open Innovation*, Oxford: Oxford University Press, pp. 3–28. https://doi.org/10.1093/acprof:oso/9780199682461.003.0001.

Chesbrough, H. W. (2003), "The logic of open innovation: managing intellectual property", *California Management Review*, Vol. 45, No. 3, pp. 33.

Chevassus-Lozza,E. and Galliano, D. (2003), "Local spillovers, firm organization and export behaviour: evidence from the French food industry", *Regional Studies*,Vol. 37, No. 2, pp. 147–158.

Choffray, J. M. and Dorey, F. (1983), *Development and Management of New Products, Concepts, Methods and Applications*, New York: McGraw-Hill.

Chowdhury, S. A., Azam, K. G. and Islam, S. (2013), "Problems and prospects of SME financing in Bangladesh", *Asian Business Review*, Vol. 2, No. 4, pp. 51–58.

Christensen, C. M., & Overdorf, M. (2000), Meeting the challenge of disruptive change. *Harvard Business Review*, 78(2), 66–77.

Christensen, J. F., Olsesen, M. H. and Kjaer, J. S. (2005), "The industrial dynamics of open innovation-evidence from the transformation of consumer electronics", *Research Policy*, Vol. 34, No. 10, pp. 1533–1549.

Chuang, L. M. (2005), "An empirical study of the construction of measuring model for organizational innovation in Taiwanese high-tech enterprises", *The Journal of American Academy of Business*, Vol. 9, No. 2, pp. 299–304.

Clancy, J. S. (2001), "Barriers to innovation in the briquetting industry in India", *Journal of Science, Technology and Society*, Vol. 6, No. 2, pp. 329–357.

Collinson, S. and Houlden, J. (2005), "Decision-making and market orientation in the internationalization process of small and medium-sized enterprises", *MIR: Management International Review*, Vol. 45, No. 4, pp. 413–436.

Cooper, J. R. (1998), "A multidimensional approach to the adoption of innovation", *Management Decision*, Vol. 36, No. 8, pp. 493–502.

Cooper, R. G. (1984), "New product strategies: what distinguishes the top performers?", *Journal of Product Innovation Management*, Vol. 2, pp. 151–64.

Cooper, R. G. (1994), "New products: the factors that drive success", *International Marketing Review*, Vol. 11, No. 1, pp. 60–76.

Corso, M., Martini, A., Paolucci, E. and Pellegrini, L. (2001), "Knowledge management in product innovation: an interpretative review", *International Journal of Management Review*, Vol. 3, No. 4, pp. 341–352.

Cortes, M. (1987), *Success in Small and Medium Scale Enterprises: The Evidence from Colombia*, New York; Oxford: Oxford University Press.

Cromie, S. (1991), "The problems experienced by young firms", *International Small Business Journal*, Vol. 9, No. 3, pp. 55–69.

Currie, D. J., A. P. Francis and J. T. Kerr. (1999), "Some general propositions about the study of spatial patterns of species richness", *Ecoscience*, Vol. 6, pp. 392–399.

Damanpour, F. (1991), Organizational innovation: a meta-analysis of effects of determinants and moderators, *Academy of Management Journal*, Vol. 34, No. 3, pp. 555–590.

Damanpour, F. and Evan, W. M. (1984), "Organizational innovation and performance: the problem of 'organizational lag'", *Administrative Science Quarterly*, Vol. 29, No. 3, pp. 392.

Damanpour, F. and Gopalakrishnan, S. (2001), "The dynamics of the product and process innovations in organizations", *Journal of Management Studies*, Vol. 38, No. 1, pp. 45–65.

Danneels, E. and Kleinschmidt, E. J. (2001), "Product innovativeness from the firm"s perspective: its dimensions and their relation with project selection and performance", *The Journal of Product Innovation Management*, Vol. 18, No. 6, pp. 357–373.

Das, K. (2008), "SMEs in India: issues and possibilities in times of globalisation", in *SME in Asia and Globalization, Research Project Report 2007*, Indonesia: Economic Resarch Institute for ASEAN and East Asia, pp. 69–97.

Das, K., Morris, S., Basant, R., Ramachandran, K. and Koshy, A. (2001), *The Growth and Transformation of Small Firms in India*, New Delhi: Oxford University Press.

Dasanayaka, S. W. S. B. (2011), "Global challenges for SMEs in Sri Lanka and Pakistan in comparative perspectives", *Business Review*, Vol. 6, No. 1, pp. 61–80.

David, F. R. (1989), "How companies define their mission", *Long Range Planning*, Vol. 22, pp. 90–97.

De Toni, A. and Nassimbeni, G. (1996), "Strategic and operational choices for small subcontracting firms- Empirical results and an interpretative mode", *International Journal of Operations and Production Management. Bradford*, Vol. 16, No. 6, pp. 41–55.

De Vos, A., Dewettinck, K. and Buyens, D. (2009), "The professional career on the right track, A study on the interaction between career self-management and organizational career management in explaining employee outcomes", *European Journal of Work and Organizational Psychology*, Vol. 18, No. 1, pp. 55–80.

Deloitte Haskins and Sells. (1989), *Management Challenge for the 1990s: The Current Education, Training and Development Debate*, Sheffield: Department of Employement, Vol. 45, pp. 135–149.

DenHartog, D. and Verburg, R. (2004), "High performance work systems, organisational culture and firm effectiveness", *Human Resource Management Journal*, Vol. 14, No. 1, pp. 55–79.

Deshpande, R., Farley, J. U. and Webster, F. E. (1993), "Corporate culture, customer orientation, and innovativeness in Japanese firms: a quadrad Analysis", *Journal of Marketing*, Vol. 57, pp. 23–37.

Deshpande, S. P. and Golhar, D. Y. (1994), "HRM practices in large and small manufacturing firms: a comparative study", *Journal of Small Business Management*, Vol. 32, No. 2, pp. 49–56.

Devanna, M. A., Fombrun, C. J. and Tichy, N. M. (1984), "A framework for strategic human resource management", *Strategic Human Resource Management*, Vol. 11, pp. 33–56.

Dewar, R. D. and Dutton, J. E. (1986), 'The adoption of radical and incremental innovations: an empirical analysis', *Management Science*, Vol. 32, No. 11, pp. 1422–1433.

Dosi, G. (1988), "Sources, procedures and microeconomic effects of innovation", *Journal of Economic Literature*, Vol. 26, No. 3, pp. 1120–1171.

Drejer, A. (2003), "Innovation and learning", *International Journal of Innovation and Learning*, Vol. 1, No. 1, pp. 8–23.

Du Plessis, M. (2007), 'The role of knowledge management in innovation', *Journal of Knowledge Management*, Vol. 11, No. 4, pp. 20–29.

Duberley, J. P. and Walley, P. (1995), "Assessing the adoption of HRM by small and medium-sized manufacturing organizations", *International Journal of HumanResource Management*, Vol. 6, No. 4, pp. 891–909.

Edoho, F. M. (2016), "Entrepreneurship paradigm in the new millennium: a critique of public policy on entrepreneurship", *Journal of Entrepreneurship in Emerging Economies*, Vol. 8, No. 1, pp. 279–294.

Edwards, T. and Delbridge, R. (2001), "Linking innovative potential to SME performance: an assessment of enterprises in industrial South Wales", in Paper for 41st European Regional Science Association Meeting, Zagreb, Croatia, Available on http://www.ersa.org/ersaconfs/ersa01/papers/full/135.pdfS (accessed May 30, 2007).

Egbetokun, A. A., Adeniyi, A. A., Siyanbola, W. O. and Olamade, O. O. (2012), "The types and intensity of innovation in developing country SMEs: evidences from a Nigerian subsectoral study", *International Journal of Learning and Intellectual Capital (IJLIC)*, Vol. 9, No. 1/2, pp. 1–16.

Elbanna, S. (2007), "The nature and practice of strategic planning in Egypt", *Strategic Change*, Vol. 16, pp. 227–243.

Engel, D., Rothgang, M. and Trettin, L. (2004), "Innovation and their impact on growth of SME – empirical evidence from craft dominated industries in Germany", Paper presented at the EARIE 2004 Conference, 2–5 September, Berlin, Germany.

Ennis, S. (1998), "Marketing planning in the smaller evolving firm: empirical evidence and reflections", *Irish Marketing Review*, Vol. 11, No. 2, pp. 49–61.

Ettlie, J. E. and Bridges, W. P. (1982), "Environmental uncertainty and organisational technology", *IEEE Transactions on Engineering Management*, Vol. 29, pp. 2–10.

European Commission. (2004), "Innovation in Europe: results for the EU, Iceland and Norway", in *Data 1998–2001*, Luxembourg: Office for Official Publications of the European Communities.

Fagerberg J.,and Verspagen B. (2009), "Innovation studies: the emerging structure of a new scientific field", *Research Policy*, Vol. 38, No.3, pp. 218–233.

Fatimah Y. A., Biswas W., Mazhar I. and Islam M. N. (2013), "Sustainable manufacturing for Indonesian small- and medium-sized enterprises (SMEs): the case of remanufactured alternators", *Journal of Remanufacturing*, Vol. 4, pp. 181–211.

Fernez Walch, S. and Romon, F. (2006), *Innovation Management from Strategy to Projects*, 2nd ed., Vuibert, French Management Review, Paris, pp. 87–103.

Fiet, J. O. (1996), "The informational basis of entrepreneurial discovery", *Small Business Economics*, Vol. 8, No. 6, pp. 419–430.

Filson A. and Lewis A. (2010), "Cultural issues in implementing changes to new product development process in a small to medium sized enterprise (SME)", *Journal of Engineering Design*, Vol. 10, pp. 97–130.

Flanagan, D. J. and Deshpande, S. P. (1996), "Top management's perceptions of changes in HRM practices after union elections in small firms", *Journal of Small Business Management*, Vol. 34, No. 4, pp. 23–34.

Fleisher, G. and Bensoussan, B. (2003), "Strategic and competitive analysis New Jersey: Prentice Hall", in Academy of Management Best Conference Paper, New York, pp. 52–67.

Foote, D. A. and Folta, T. B. (2002), "Temporary workers as real options", *Human Resource Management Review*, Vol. 12, pp. 579–597.

Forsman, H. and Annala, U. (2011), "Small enterprises as innovators: shift from a low performer to a high performer", *International Journal of Technology Management (IJTM)*, Vol. 56, No. 2/3/4, pp. 154–171.

Fossen, R., Rothstein, H. and Korn, H. (2006), "Thirty-five years of strategic planning and performance research: a meta-analysis", in Academy of Management Best Conference Paper BPS: M1, New York, pp. 77–91.

Francis, D., & Bessant, J. (2005), Targeting innovation and implications for capability development. *Technovation*, Vol. 25, No. 3, pp. 171–183.

Freel, M. S. (2005), "Patterns of innovation and skills in small firms", *Technovation*, Vol. 25, No. 2, pp. 123–134.

Freeman, J., Lawley, M. and Styles, C. (2010), "Does firm location influence the export performance of Australian SMEs?", in 2010 Conference of the Australian and New Zealand Marketing Academy (ANZMAC), Christchurch, New Zealand.

Frenkel, A. (2001), "Why high technology firms choose to locate in or near metropolitan areas", *Urban Studies*,Vol. 38, No. 7, pp. 1083–1101.

Frenz, M., Michie, J. and Oughton, C. (2003), "Regional dimension of innovation: results from the third community innovation survey", in International Workshop Empirical Studies on Innovation in Europe, Faculty of Economics, Italy: University of Urbino, pp. 733–745.

Fritsch M. and Meschede, M. (2001), "Product innovation, process innovation, and size", *Review of Industrial Organization*, Vol. 19, No. 3, pp. 335–350.

Frost, F. (2003), "The use of strategic tools by small and medium-sized enterprises: an Australasian study", *Strategic Change*, Vol. 12, pp. 49–62.

Fugate, M. and Kinicki, A. J. (2008), "A dispositional approach to employability: development of a measure and test of implications for employee reactions to organizational change", *Journal of Occupational & Organizational Psychology*, Vol. 81, No. 3, pp. 503–527.

Garcia, R. and Calantone, R. (2002), A critical look at technological innovation typology and innovativeness terminology: a literature review. *Journal of Product Innovation Management*, Vol. 19, No. 2, pp. 110–132.

Gardiner, P., & Rothwell, R. (1985), Invention, innovation, re-innovation and the role of the user: a case study of British hovercraft development. *Technovation*, Vol. 3, pp. 167–186.

Gatignon, H., Tushman, M. L., Smith, W. and Anderson, P. (2002), "A structural approach to assessing innovation: construct development of innovation locus, type and characteristics", *Management Science*, Vol. 48, No. 9, pp. 1103–1122.

Gedam, R. (2011), "Towards reducing India's power sector aggregate technical and commercial losses", http://www.spml.co.in/mediamroom/Electrical%20Monitor%20Artical%20%20Dr%20%20Ratnaker%20Gedam.pdf.

Geroski, P. (1994), *Market Structure, Corporate Performance and Innovative Activity*, Oxford: Clarendon Press.

Ghobadian, A. and Gallear, D. N. (1996), "Total quality management in SMEs", *Omega International Journal Managementt Science*, Vol. 24, No. 1, pp. 83–106.

Glaister, K. and Falshaw, R. (1999), "Strategic planning: still going strong?", *Long Range Planning*, Vol. 32, No. 1, pp. 107–116.

Global Innovation Index Report. (2015), Available online at http://www.wipo.int/export/sites/ www/freepublications/en/economics/gii/gii_2015.pdf

Gloet, M. and Terziovski, M. (2004), 'Exploring the relationship between knowledge management practices and innovation performance', *Journal of Manufacturing Technology Management*, Vol. 15, No. 5, pp. 402–409.

Godin B. (2006), "The linear model of innovation: the historical construction of an analytical framework", *Science Technology and Human Values*, Vol. 31, No. 6, pp. 631–667.

Gorzen-Mitka, I. (2013), Risk identification tools: polish MSMES companies practices, *Problems of Management in the 21st Century*, Vol. 7/2013, pp. 6–11.

Greco, M., Grimaldi, M. and Cricelli, L. (2015), "Open innovation actions and innovation performance: a literature review of European empirical evidence", *European Journal of Innovation Management*, Vol. 18 No. 2, pp. 150–171.

Greening, D. W., Barringer, B. R. and Macy, G. A. (1996), "Qualitative study of managerial challenges facing small business geographic expansion", *Journal of Business Venturing*, Vol. 11, pp. 233–256.

Greenley, G. (1994), "Strategic planning and company performance: an appraisal of empirical evidence", *Scandinavian Journal of Management*, Vol. 10, No. 4, pp. 383–396.

Grossman, G. and Helpman, E. (1994), 'Endogenous innovation in the theory of growth', *Journal of Economic Perspectives*, Vol. 8, No. 1, pp. 23–44.

Guest, D. E. (1997), "Human resource management and performance: a review and research agenda", *International Journal of Human Resource Management*, Vol. 8, No. 3, pp. 263–276.

Gupta, M. and Cawthon, G. (1996), "Managerial implications of flexible manufacturing for small/medium-sized enterprises", *Technovation*, Vol. 16, No. 2, pp. 77–83.

Gupta, U. and Tannenbaum, J. A. (1989), "Enterprise: labor shortages force changes at small firms", *Wall Street Journal, B-2*, Vol. 37, pp. 172–187.

Guzzo, R. A., Jette, R. D. and Katzell, R. A. (1985), "The effects of psychologically based intervention programs on worker productivity: a meta-analysis", *Personnel Psychology*, Vol. 38, pp. 275–291.

Hall, D. T. (2002), "*Careers in and out of Organizations*", Thousand Oaks: Sage Publications, Vol. 36, pp. 105–121.

Hallier, J. (2009), "Rhetoric but whose reality? The influence of employability messages on employee mobility tactics and work group identification", *International Journal of Human Resource Management*, Vol. 20, No. 4, pp. 846–868.

Hankinson, A. (1991), "Small business: management and performance", *Academy of Management Journal*, Vol. 21, pp. 73–87.

Harris, R. I. D. and Robinson, C. (2002), "Research Project on DTI Industrial Support Policies", *DTI Final Report Ref: SEC Research 01ITT No. SEC01*, London: Department of International Trade.

Hatcher, L. (1994), *A Step by Step Approach to Using SAS for Factor Analysis and Structural Equation Modelling*. Cary: SAS Institute Inc.

Hausman, A. (2005), "Innovativeness among small businesses: theory and propositions for future research", *Industrial Marketing Management*, Vol. 34, No. 8, pp. 773–782.

Hayashi, M. (2002), "The role of subcontracting in SME development in Indonesia: micro-level evidence from the metal working and machinery industry", *Journal of Asian Economics*, Vol. 13, No. 1, pp. 1–26.

Hayland, P. W. (2004), "Innovation and enhancement of enterprise capabilities", *IJMT*, Vol. 3, No. 1, pp. 35–46.

Heathfield, P. (1997), "SME business leaders need powerful on-board computers", *Industrial Management & Data Systems*, Vol. 97, No. 6, pp. 233–235.

Hebert, R. F. and Link, A. N. (1989), "In search of the meaning of entrepreneurship", *Small Business Economics*, Vol. 1, No. 1, pp. 39–49.

Henderson, R. M., and Clark, K. B. (1990), Architectural innovation: the reconfiguration of existing product technologies and the failure of established firms. *Administrative Science Quarterly*, Vol. 35, No. 1, pp. 9–30.

Heneman, H. G. and Berkley, R. A. (1999), "Applicant attraction practices and outcomes among small businesses", *Journal of Small Business Management*, Vol. 14, pp. 53–74.

Herr, E. L. and Cramer, S. G. (1996), "Information in career guidance and counseling through the lifespan: systematic approaches", *The Learning Organization*, Vol. 7, No. 1, pp. 5–12.

Heunks, F. (1998), 'Innovation, creativity and success', *Small Business Economics*, Vol. 10, pp. 263–272.

Hornsby, J. S. and D. F. Kuratko, (1990), "Human resource management in small business: critical issues for the 1990's", *Journal of Small Business Management*, Vol. 28, No. 3, pp. 9–18.

Hsueh, L. and Tu, Y. (2004), 'Innovation and the operational performance of newly established small and medium enterprises in Taiwan', *Small Business Economics*, Vol. 23, pp. 99–113.

Huang, C., Amorim, C., Spinoglio, M., Gouveia, B. and Medina, A. (2004), "Organization, programme and structure: an analysis of the Chinese innovation policy framework", *R&D Management 34*, Vol. 4, pp. 167–181.

Huang, S. C. (2008), "Efficient industrial technology policy, high government industrial R&D expenditure: does one require the other?", *International Journal of Technology, Policy and Management*, Vol. 8, No. 3, pp. 211–236.

Huang, X. and A. Brown, (1999), "An analysis and classification of problems in small business", *International Small Business Journal*, Vol.18, No. 1, pp. 73–85.

Hughes, A. (2001), 'Innovation and business performance. Small entrepreneurial firms in the UK and the EU', *New Economy*, Vol. 8, No. 3, pp. 157–163.

Huiban, J. P. and Bouhsina, Z. (1998), "Innovation and the quality of labour factor: an empirical investigation in the French food industry", *Small Business Economics*, Vol.10, pp. 389–400.

Huselid, M. A. (1995), "The impact of human resource management practices on turnover, productivity, and corporate financial performance", *Academy of Management Journal*, Vol. 38, No. 3, pp. 635–670.

Hussain, I., Farooq, Z. and Akhtar, W. (2011), "SMEs development and failure avoidance in developing countries through public private partnership", *African Journal of Business Management*, Vol. 6, No. 4, pp. 1581–1589.

Hyytinen, A. and Toivanen, O. (2005), "Do financial constraints hold back innovation and growth? evidence on the role of public policy", *Research Policy*, Vol. 34, No. 9, pp. 1385–1403.

Inkson, K. and King, Z. (2010), "Contested terrain in careers: a psychological contract model", *Human Relations*, Vol. 64, No. 1, pp. 37–57.

Jahnshahi,A., Nawaser,K., Paghaleh,M. J., Mohammad,S. and Khaksar, S. (2011), "The role of government policy and the growth of entrepreneurship in the micro, small (&) medium-sized enterprises in india: an overview", *Australian Journal of Basic and Applied Sciences*, Vol. 5, No. 3, pp. 1563–1571.

James, A., Gee, S., Love, J., Roper, S. and Willis, J. (2014), "Small firm-large firm relationships and the implications for small firm innovation: what do we know?", in DRUID Society Conference 2014, CBS, Copenhagen.

Jennings, R. and Cox, C. (1995), "The foundations of success: the development and characteristics of British entrepreneurs and intrapreneurs", *Leadership & Organization Development Journal*, Vol. 16, No. 7, pp. 4–9.

Jeong, K. and Phillips, D. T. (2001), "Operational efficiency and effectiveness measurement", *International Journal of Operations and Production Management*, Vol. 21, No. 11, pp. 1404–1416.

Jones, M. and Jain, R. (2002), "Technology transfer for SMEs: challenges and barriers", *International Journal on Technology Transfer and Commercialization*, Vol. 1, No. 1/2, pp. 146–162.

Jordan, J., Lowe, J. and Taylor, P. (1998), "Strategy and financial policy in UK small firms", *Journal of Business Finance & Accounting*, Vol. 25, No. 1/2, pp. 1–27.

Kache, F., Bettermann, L. and Magerle, R. (2011), "Gaining competitive advantage through more effective direct material sourcing", http://www.accenture.com /SiteCollectionDocuments/PDF/Accenture-Gaining-Competitive-Advantage -through-More-Effective-Direct-Material-Sourcing.pdf.

Kalra, S. C. (2009), "SMEs in India: the challenges ahead", http://articles.economictimes.indiatimes.com/2009-02-03/news/28468202_1_indian-sme-sector-sme-segment-sme-units.

Kaman, V., McCarthy, A. M., Gulbro, R. D. and Tucker, M. L. (2001), "Bureaucratic and high commitment human resource practices in small service firms", *Human Resource Planning*, Vol. 24, No. 1, pp. 33–44.

Kaplinsky, R. and Manning, C. (1998), "Concentration policy and the role of small and medium-sized enterprises in South Africa's industrial development", *Journal of Development Studies*, Vol. 35, No. 1, pp. 139–150.

Karagozoglou, N. and Lindell, M. (1998), "Internationalization and small and medium sized technology-based firms: an exploratory study", *Journal of Small Business Management*, Vol. 36, No. 1, pp. 44–59.

Kaushik, S. P. and Kaur, J. (2001), "Evaluation of industrial politics and infrastructure facilities in national capital sub-region, Haryana", *Indian Journal of Geography and Environment*, Vol. 12, pp. 31–39.

Keeble, D. (1997), "Small firms, innovation and regional development in Britain in 1990s", *Regional Studies*, Vol. 31, No. 3, pp. 281–293.

Keefe, L. M. (2004), "What is the meaning of marketing", *Marketing News*, Vol. 38, No. 15, pp. 17–18.

Keizer, J., Dijkstra, L. and Halman, J. (2002), 'Explaining innovative efforts of SMEs. An exploratory survey among SMEs in the mechanical and electrical engineering sector in Netherlands', *Technovation*, Vol. 22, No. 1, pp. 1–13.

Khayyat, N. T. and Lee, J. D. (2015), "A measure of technological capabilities for developing countries", *Technological Forecasting & Social Change*, Vol. 92, pp. 210–223.

Kim, L. (1988), "Entrepreneurship and innovation in a rapidly developing country", *Journal of Development Planning*, Vol. 8, pp. 183–194.

Kim, S. (1991), "Product performance as a unifying theme in concurrent design concepts", *Robotics and Computer-Integrated Manufacturing*, Vol. 8, No. 2, pp. 121–126.

Knight, K. E. (1967), "A descriptive model of the intra-firm innovation process", *The Journal of Business*, Vol. 40, pp. 478.

Koch, M. J. and McGrath, R. G. (1996), "Improving labor productivity: human resource management policies do matter", *Strategic Management Journal*, Vol. 17, pp. 335–354.

Kor, Y., Mesko, A., (2013), "Dynamic managerial capabilities: configuration and orchestration of top executives" capabilities and the firm's dominant logic", *Strategic Management Journal*, Vol. 23, pp. 233–244.

Kotler, P. (1977), "From sales obsession to marketing effectiveness", *Harvard Business Review*, Vol. 55, pp. 67–75.

Kotler, P. (1999), *Marketing Management*, 10th ed., Glencoe: IL Free Press.

Koufopoulos, D., Logoudis, I. and Pastra, A. (2005), "Planning practices in the Greek ocean shipping industry", *European Business Review*, Vol. 17, No. 2, pp. 151–176.

Krishnaswamy, T. S. (2009), *MSME Financing: How to Make It Easy for an Entrepreneur?*, Mumbai: State Bank of India, Corporate Centre.

Kristiansen S. (2003), "Small-scale business in rural java: involution or innovation?", *Journal of Entrepreneurship*, Vol. 12, No. 1, pp. 21–41.

LaBarbera, P. A. and Rosenberg, S. A. (1989), *Marketing Research and Small Entrepreneurial Enterprises*, Chicago: Published by the Office for Entrepreneurial Studies, Vol. 12, No. 1, pp. 233–244.

Lado, A. A., Wilson, M. C. (1994), "Human resource systems and sustained competitive advantage: a competency-based perspective", *Academy of Management Journal*, Vol. 19, No. 4, pp. 699–727.

Laforet, S. (2013), Organizational innovation outcomes in SMEs: effects of age, size, and sector, *Journal of World Business*, Vol. 48, pp. 490–502.

Lagrosen, S. (2005), 'Customer involvement in new product development, a relationship marketing perspective', *European Journal of Innovation*, Vol. 8, No. 4, pp. 424–436.

Lahiri, R. (2012), "Problems and prospects of micro, small and medium enterprises (MSMEs) in India in the era of globalization", in International Conference at Royal Thimphu College, Bhutan.

Lall, S. (1992), "Technological capabilities and industrialization", *World Development*, Vol. 20, No. 2, pp. 165–186.

Landabaso, M. (2000), "EU policy on innovation and regional development", *Knowledge, Innovation and Economic Growth: The Theory and Practice of Learning Regions*. Cheltenham: Edward Elgar, Vol. 12, pp. 423–441.

Lange, P. (2011), "Africa - internet, broadband and digital media statistics 69", Retrieved from https://www.budde.com.au/Research/Africa-Internet-Broadband-and-DigitalMedia-Statistics-tables-only.html.

Larson, C. M. and Clute, R. C. (1979), "The failure syndrome", *American Journal of Small Business*, Vol. 4, No. 2, pp, 36–42.

Larsson, E., Hedelin, L. and Garling, T. (2003), "Influence of expert advice on expansion goals of small businesses in Rural Sweden", *Journal of Small Business Management*, Vol. 41, No. 2, pp. 205–221.

Lazarova, M. and Taylor, S. (2009), "Boundaryless careers, social capital, and knowledge management: implications for organizational performance", *Journal of Organizational Behavior*, Vol. 30, No. 1, pp. 119–139.

Lee, J. (1995), "Small firm's innovation in two technological settings", *Research Policy*, Vol. 24, No. 3, pp. 391–401.

Lee, S. K. and Lee, S. H. (2010), "The activation of technology finance through support for small and medium-sized enterprises in Korea", *International Journal of Business and Management*, Vol. 5, No. 4, pp. 75–79.

Lehtimaki, A. (1991), 'Management of the innovation process in small companies in Finland', *IEEE Transactions on Engineering Management*, Vol. 38, No. 2, pp. 120–126.

Leonard, D. A. and Rayport, J. F. (1997), "Spark innovation through empathic design", *Harvard Business Review*, Vol. 75, No. 6, pp. 102–113.

Leppard, J. and McDonald, M. (1987), "A re-appraisal of the role of marketing planning", *Journal of Marketing Management*, Vol. 3, No. 2, pp. 159–171.

Li, X. and Wu, G. (2010), "In-house R&D, technology purchase and innovation: empirical evidences from Chinese hi-tech industries, 1995–2004", *International Journal of Technology Management (IJTM)*, Vol. 51, No. 2/3/4, pp. 217–238.

Lilien, G. L., Kotler, P. and Moorthy, K. S. (1992), "Marketing models", in *Long Range Planning Englewood Cliffs*, New Jersey: Prentice Hall International, Vol. 62, pp. 103–122.

Liyanage, S. (2003), "Technology and innovation management learning in the knowledge economy", *Journal of Management Development*, Vol. 22, No. 7, pp. 579–602.

Llorens-Montes, F. J. and Garcia-Morales, V. J. (2004), "The influence on personal mastery, organisational learning and performance of the level of innovation: adaptive organization versus innovator organization", *International Journal of Innovation and Learning*, Vol. 1, No. 2, pp. 101–114.

Lumiste, R, Lumiste, R. and Kilvits, K. (2004), 'Estonian manufacturing SMEs innovation strategies and development of innovation networks', in 13th Nordic Conference on Small Business Research, Norway: University of Oslo.

Lundvall, B. K. and Nielsen, P. (2007), 'Knowledge management and innovation performance', *International Journal of Manpower*, Vol. 28, Nos. 3/4, pp. 207–223.

MacGregor, R. and Varazalic, L. (2005), "A basic model of electronic commerce adoption barriers, a study of regional small businesses in Sweden and Australia", *Journal of Small Business and Enterprise Development*, Vol. 12, No. 4, pp. 510–527.

Malmberg, A., Malmberg, B. and Lundequist, P. (2000), "Agglomeration and firm performance: economies of scale, localization and urbanization among swedish export firms", *Environment and Planning A*, Vol. 32, No. 2, pp. 305–321.

Mambula, C. (2002), "Perceptions of SME growth constraints in Nigeria", *Journal of Small Business Management*, Vol. 40, No. 1, pp. 58–65.

Man T. W. Y., Lau T and Snape E. (2012), "Entrepreneurial competencies and the performance of small and medium enterprises: an investigation through a framework of competitiveness", *Journal of Small Business & Entrepreneurship*, Vol. 8, pp. 243–288.

Mansfield, E. (1968), *The Economics of Technological Change*, New York: W. W. Norton, pp. 699–700.

Mansfield, E. (1971), *Research and Innovation in the Modern Corporation*, W. W. Norton.

Mansfield, E. and Jeong-Yeon, L. (1996), "The modem university: contributor to industrial innovation and recipient of industrial R&D support", *Research Policy*, Vol. 25, pp. 1047–1058.

Marcati, A., Guido, G. and Peluso, A. M. (1998), "What is marketing for SME entrepreneurs? The need to market the marketing approach", *Journal of Marketing Management*, Vol. 12, pp. 161–173.

Marjanova, T. J. (2008), "Marketing knowledge and strategy for SMEs: can they live without it?", *SAM Advanced Management Journal*, Vol. 43, No. 2, pp. 24–48.

Masurel, E., Montfort, K. and Lentink, R. (2010), "Innovation and diffusion in small firms: theory and evidence", *Small Business Economics*, Vol. 6 No. 4, pp. 327–347.

Mathew, J. and Vijay, M. (2012), "Technology business incubators: a perspective for the emerging economies", *Indian Journal of Business Management*, Vol. 14, No. 3, pp. 358–365.

Matthews, C. H. and Scott, S. G. (1995), "Uncertainty and planning in small and entrepreneurial firms: an empirical assessment", *Journal of Small Business Management*, Vol. 33, No. 4, pp. 34–52.

Matusik, S. and Hill, C. (1998), "The utilization of contingent work, knowledge creation, and competitive advantage", *Academy of Management Review*, Vol. 23, pp. 680–697.

Maupin, A. J. and Stauffer, A. L. (2000), "A design tool to help small manufacturers reengineer a product family", in Proceedings of 2000 ASME Design Engineering Technical Conference, Baltimore.

Mavondo, F. T., Chimhanzi, J. and Stewart, J. (2005), 'Learning orientation and market orientation: relationship with innovation, human resource practices and performance', *European Journal of Marketing*, Vol. 39, No. 11, pp. 1235–1263.

May, K. (1997), "Work in the 21st century: understanding the needs of small businesses", *Industrial and Organizational Psychologist*, Vol. 35, No. 1, pp. 94–97.

Mazzarol, T. (2004), "Strategic management of S mall firms: a proposed framework for entrepreneurial ventures", in 17th Annual SEAANZ Conference - Entrepreneurship as the Way of the Future, Brisbane, Queensland.

McEvoy, G. M., (1984), "Small business personnel practices", *Journal of Small Business Management*, Vol. 22, No. 4, pp. 1–8.

McGregor, H. (2011), "Human resource personnel management", *West Publishing Company, St. Paul Ministry*, Vol. 7, pp. 85–92.

McKone, K. E., Schroeder, R. G. and Cua, K. O. (1999), "Total productive maintenance: contextual view", *Journal of Operations Management*, Vol. 17, No. 2, pp. 123–134.

Milanzi, M. A. (2012), "Export barrier perceptions in Tanzania: the influence of social networks", *Journal of African Business*, Vol. 13, No. 1, pp. 29–39.

Ministry of MSME. (2015), *Annual Report 2009-10, Ministry of Micro, Small and Medium Enterprises*, New Delhi: Government of India.

Moore, B. (1995), *What Differentiates Innovative Small Firms?, Innovation Initiative Paper No. 4*, England and London: ESRC Centre for Business Research, University of Cambridge.

Mori, N. and Munisi, G. (2012), "The role of the internet in overcoming information barriers: implications for exporting SMEs of the East African community", *Journal of Entrepreneurship, Management and Innovation*, Vol. 8, No. 2, pp. 60–77.

Morris, S. and Das, P. (2001), "The growth and transformation of small firms in India", *Journal of Small Business Management*, Vol. 4, pp. 43–49.

Mosey, S., Clare, J. and Woodcock, D. (2002), 'Innovation decision making in British manufacturing SMEs', *Integrated Manufacturing Systems*, Vol. 13, No. 3, pp. 176–183.

Mothe, C. and Thi, T. U. N. (2010), "The link between non-technological innovations and technological innovation", *European Journal of Innovation Management*, Vol. 13, No. 3, pp. 313–332.

Motwani, J. G., Jiang, J. J. and Kumar, A. (1998), "A comparative analysis of manufacturing practices of small vs. large West Michigan organizations", *Industrial Management & Data Systems*, Vol. 98, No. 1, pp. 8–11.

Munster, M. (2011), "What does open innovation imply for high-tech SMEs?, An Empirical Study and Analysis on Motives, Challenges and Consequences", *School of Management and Governance,* Netherlands: University of Twente, pp. 1–57.

Nanda, T. and Singh, T. P. (2009), "An assessment of the technology innovation initiatives in the Indian small manufacturing industry", *International Journal of Technology*, Vol. 9, No. 2, pp. 173–207.

Narver, J. C. and Slater, S. F. (1990), "The effect of market orientation on business profitability", *Journal of Marketing*, Vol. 54, No. 4, pp. 20–35.

Nath, R. and Singh, G. (2010), "Creating competitive SMEs", http:www.cii.in/webcams/upload/creating%competing%20SMEs.pdf.

Nauwelaers, C. and Wintjes, R. (2002), "Innovating SMEs and regions: the need for policy intelligence and interactive policies", *Technology Analysis & Strategic Management*, Vol. 14, No. 2, pp. 223–242.

Nelson R. E and M. F. Mwaura, (1997), "Growth strategies of medium-sizes firms in Africa", *The Journal of Entrepreneurship*, Vol. 6, No.1, pp. 53–73.

Nganga, S. I. (2011), "Collective efficiency and its effects on infrastructure planning and development for small manufacturing enterprises in Kenya", *International Journal of Business and Public Management*, Vol. 1, No. 1, pp. 75–84.

Nieto, M. J. and Santamaria, L. (2010), "Technological collaboration: bridging the innovation gap between small and large firms", *Journal of Small Business Management*, Vol. 48, No. 1, pp. 44–69.

Nkuah, J. K., Tanyeh, J. P. and Gaeten, K. (2013), "Financing small and medium enterprises (SMEs) in Ghana: challenges and determinants in accessing bank credit", *International Journal of Research in Social Sciences*, Vol. 2, No. 3, pp. 12–25.

Nonaka, I. and Takeuchi, H. (1995), *"The Knowledge Creating Company"*, Oxford: Oxford University Press.

Nooteboom, B. (1999), "Innovation and inter-firm linkages: new implications for policy", *Research Policy*, Vol. 28, pp. 793–805.

Nunnally, J. C. (1978), *Psychometric Theory*, New York: McGraw-Hill.

O"Gorman, C. (1997), "Success strategies in high growth small and medium-sized enterprises", in Jones-Evans, D. and Klofsten, M. (Eds.), *Technology, Innovation and Enterprise*, The European Experience, Palgrave Macmillan; 1997th edition, London, U.K. pp. 1–423.

Oakey, R. (1991), "Innovation and the management of marketing in high technology small firms", *Journal of Marketing Management*, Vol.7, No. 4, pp. 343–356.

OECD. (2000), *Small and Medium-sized Enterprises: Local Strength, Global Reach*, Paris: Organisation for Economic Co-operation and Development.

OECD/Eurostat. (2005), *Oslo Manual: Guidelines for Collecting and Interpreting Innovation Data*, 3rd ed., Paris: Organisation for Economic Co-operation and Development: Statistical Office of the European Communities.

Ojala, M., Vilpola, I. and Kouri, I. (2006), "Risks and risk management in ERP Project - cases in SME Context", *Business Information Systems*, Vol. 13, pp. 134–139.

Okreglicka, M. (2014), "Adoption and Use of ICT as a Factor of Development of Small and Medium-sized Enterprises in Poland", *Entrepreneurship and Management*, Vol. 15, No. 7, pp. 393–405.

Olav, S. and Leppälahti, A. (1997), 'Innovation, firm profitability and growth', *Paper Provided by the Step Group, Studies in Technology, Innovation and Economic Policy in Step Report Series No. 1997/01.*

Olawale, F. and Garwe, D. (2010), "Obstacles to the growth of new SMEs in South Africa: a principal component analysis approach", *African Journal of Business Management*, Vol. 4, No. 5, pp. 729–738.

Oluleye, F. A. and Oyetayo, O. (2010), "Raw materials development and utilisation in Nigeria: promoting effective linkage between R&D and SMEs for economic growth and development", *Journal of Management and Corporate Governance*, Vol. 2, pp. 41–54.

Ominde, W. (1964), *Education Commission Report–1965*, Nairobi: Kenya: Government Printers.

Oosterbeek, H., (1998), "Unravelling supply and demand factors in work-related training", *Oxford EconomicPapers*, Vol. 50, pp. 266–283.

Pallant, J. (2005), *SPSS Survival Manual: A Step by Step Guide to Data Analysis Using SPSS for Window*, New South Wales, Australia: Allen and Unwin.

Park, H. S (2017), "Technology convergence, open innovation and dynamic economy", *Journal of Open Innovation*, Vol. 3, pp. 1–13. (Chapter 2)

Parr, J. B. (2002), "Agglomeration economies: ambiguities and confusion", *Environment and Planning A*, Vol. 34, No. 4, pp. 717–731.

Pasadilla, G. O. (2010), *Financial Crisis, Trade Finance, and SMEs: Case of Central Asia*, Japan: Asian Development Bank Institute.

Peres,W. and Stumpo, G. (2000), "Small and medium-sized manufacturing enterprises in Latin America and the Caribbean under the new economic model", *World Development*, Vol. 28, No. 9, pp. 1643–1655.

Peters, T. and Waterman, R. (1982), *In Search of Excellence*, New York: Harper and Row.

Pfeffer, J., (1998), "Seven practices of successful organizations", *California Management Review*, Vol. 40, No. 2, pp. 96–124.

Pich, M. T., Loch, C. H. and De Meyer, A. (2002), "On uncertainty, ambiguity, and complexity in project management", *Management Science*, Vol. 48, No. 8, pp. 1008–1023.

Polverari, L., McMaster, I., Gross, F., Bachtler, J., Ferry, M. and Yuill, D. (2006), "Strategic Planning for Structural Funds 2007–2011: A Review of Strategies and Programmes", European Policies Research Centre,Glasgow: University of Strathclyde.

Ponmani, R. (2011), "Infrastructure and SMEs development in selected Asian countries", *Asian Journal of Research in Social Science & Humanities*, Vol. 1, No. 4, pp. 465–473.

Prajogo, D. I. and Sohal, A. S. (2004), 'Transitioning from total quality management to total innovation management, an Australian case', *International Journal of Quality & Reliability Management*, Vol. 21, No. 8, pp. 861–875.

Radas, S. and Bozic, L. (2009), "The antecedents of SME innovativeness in an emerging transition economy", *Technovation*, Vol. 29, pp. 438–450.

Rakowski W., Andersen M.R. and Stoddard A.M. (1997), "Confirmatory analysis of opinions regarding the pros and cons of mammography", *Health Psychology*, Vol. 16, No. 2, pp. 433–444.

Ramanujam, V. and Varadarajan, P. (1989), "Research on corporate diversification: a synthesis", *Strategic Management Journal*, Vol. 10, No. 6, pp. 523–551.

Ramanujam, V., Venkatraman, N. and Camillus, J. (1986), "Multi-objective assessment of effectiveness of strategic planning: a discriminant analysis approach", *Academy of Management Review*, Vol. 29, No. 2, pp. 347–372.

Regev, H. (1998), 'Innovation, skilled labor, technology and performance in Israeli industrial firms', *Economics of Innovation and New Technology*, Vol. 5, pp. 301–323.

Reinikka, R. and Svensson, J. (2001), "Confronting competition: investment, profit and risk", in Uganda Recovery: The Role of Farms, Firms, and Government, Kampala.

RNCOS Industry Research Solutions. (2010), *Indian Power Sector Analysis*. http://www.rncos.com/Market-Analysis-Reports/Indian-Power-Sector-Analysis-IM114.htm.

Roberts, E. B. (2007), "Managing invention and innovation", *Research-Technology Management*, Vol. 50, No. 1, pp. 35–54.

Rogers, E. M. (2004), *Diffusion of Innovations*, New York: Free Press.

Rogers, E. M. (1995), *Diffusion of Innovations*, 4th ed., New York: Free Press.

Romano, C. and Ratnatunga, J. (1995), "The role of marketing: it's impact on small enterprise research", *European Journal of Marketing*, Vol. 29, No. 7, pp. 9–30.

Romer, P. (1994), "The origins of endogenous growth", *Journal of Economic Perspective*, Vol. 8, No. 1, pp. 3–22.

Romijn H. (2001), "Technology support for small-scale industry in developing Countries: a review of concepts and project practices", *Oxford Development Studies*, Vol. 29, No. 1, pp. 57–76.

Roper, S. (1997), "Product innovation and small business growth: a comparison of strategies of German, UK and Irish companies", *Small Business Economics*, Vol. 9, No. 6, pp. 523–537.

Roper, S., Love, J., Dunlop, S., Ashcroft, B., Hofmann, H. and Vogler-Ludwig, K. (1996), *Product Innovation and Development in UK, German and Irish Manufacturing*, Kingston, Ontario, Canada: Northern Irelan Economic Research Centre, Queens University.

Saaty, T. L. (1980), *The Analytic Hierarchy Process*, New York: McGraw-Hill Book Co.

Saaty, T. L. (1994), *Fundamentals of Decision Making*, Pittsburgh: RWS Publications.

Saleh, A. S. and Ndubisi N. O. (2006), "An evaluation of SME development in Malaysia", *International Review of Business Research Papers*, Vol. 2, No. 1, pp. 1–14.

Scandura, C. A., Gitlow, H. and Yau, S. C, (1996), "Mission statements in service and industrial corporations", *International Journal of Quality Science*, Vol. 1, pp. 48–61.

Schlogl, H. (2004), "Small and medium enterprises: seizing the potential", *Organizational for Economic Cooperation and Development, OECD Observer*, Vol. 1, No. 243, pp. 46–48, May 2004.

Schreiber, J. B., Nora, A., Stage, F. K., Barlow, E. A. and King, J. (2006), "Reporting structural equation modeling and confirmatory factor analysis results: a review", *The Journal of Educational Research*, Vol. 99, No. 6, pp. 323–337.

Schumpeter J. A. (1934), "The theory of economic development an inquiry into profits, capital, credit, interest and the business cycle", in Conference of Business Investments, Harvard University, Cambridge.

Schumpeter, J. A. (1939), *Business Cycle: A Theoretical, Historical, and Statistical Analysis of the Capitalist Process*, New York: McGraw Hill Book Company Inc.

Schumpeter, J. A. (1942), *Capitalisme, Socialisme and Democracy* (J. M. Tremblay, Trans.) Routledge, Taylor and Francis Group, London, U.K.

Schwab, K. (2017), *"The Fourth Industrial Revolution"*, New York: Crown Business (Chapter 2).

Scott, M., Roberts, I., Holroyd G. and Sawbridge, D. (1986), "Management and industrial relations in small firms", *Department of Employment Research Paper*, Vol. 5, pp. 70–81.

Scott, M., Roberts, I., Holroyd, G. and Sawbridge, D. (1989), "Management and industrial relations in small firms", *Employee Relations*, Vol. 6, No. 5, pp. 21–24.

Shah, R. and Goldstein, S. (2006), "Use of structural equation modeling in operations management: looking back and forward", *Journal of Operations Management*, Vol. 24, No. 2, pp. 148–169.

Shang, S. S. C., Wu, S. H. and Yao, C. Y. (2010), "A dynamic innovation model for managing capabilities of continuous innovation", *International Journal of Technology Management (IJTM)*, Vol. 51(2/3/4), pp. 519–528.

Sharif, N., Baark, E. and Lau, A. K. W. (2012), "Innovation activities, sources of innovation and R&D cooperation: evidence from firms in Hong Kong and Guangdong Province, China", *International Journal of Technology Management (IJTM)*, Vol. 59, No. 3/4, pp. 203–234.

Shefer, D. and Frenkel, A. (2005), 'R&D, firm size and innovation: an empirical analysis', *Technovation*, Vol. 25, pp. 25–32.

Sheppeck, M. and Militello, J. (2000), "Strategic HR configurations and organizational performance", *Human Resource Management*, Vol. 39, No. 1, pp. 5–16.

Shim, M. (2010), "Factor influencing child welfare employee's turnover: focusing on organizational culture and climate", *Children and Youth Service Review*, Vol. 32, pp. 847–856.

Shiralashetti, A. S. (2012), "Prospects and problems of MSMEs in India: a study", *International Journal of in Multidisciplinary and Academic Research (SSIJMAR)*, Vol. 1, No. 2, pp. 1–7.

Siebert, S. W. and J. Addison, (1991), "Internal labour markets: causes and consequences", *Oxford Review of Economic Policy*, Vol. 7, pp. 76–92.

Sikka, P. (1999), "Technological innovation by SME"s in India", *Technovation*, Vol. 19, No. 5, pp. 317–321.

Simon, H. A. (1996), *The Sciences of the Artificial*, 3rd ed., MIT Press: Cambridge, MA, Vol. 19, pp. 305–330.

Singh, B. (2012), "Identifying critical barriers in the growth of Indian micro, small and medium enterprises (MSMEs)", *International Journal of Business, Competition and Growth*, Vol. 2, No. 1, pp. 84–105.

Singh, B., Mathews, J., Mullineux, G. and medland, T. (2009), "Product development in manufacturing SMEs: current state, challenges and relevant supportive techniques", in International Conference on Engineering Design, Stanford University, Stanford.

Sirmon, D. and Hitt, M. (2009), "Contingencies within dynamic managerial capabilities: interdependent effects of resources investment and deployment on firm performance", *Strategic Management Journal*, Vol. 30, pp. 1375–1394.

Smith, D., (1990), "Small is beautiful, but difficult: towards cost-effective research for small business", *Journal of The Market Research Society*, Vol. 32, No. 1, pp. 37–60.

Sobanke, V., Adegbite, S., Ilori, M. and Egbetokun, A. (2014), "Determinants of technological capability of firms in a developing Country", *Procedia Engineering*, Vol. 69, pp. 991–1000.

Soderquist, K., Chanaron, J. J. and Motwani, J. (1997), 'Managing innovation in French small and medium-sized enterprises: an empirical study', *Benchmarking for Quality Management and Technology*, Vol. 4, No. 4, pp. 259–272.

Sonia and Kansal, R. (2009), "Globalization and its impact on small scale industries in India", *International Journal of Technology Management*, Vol. 1, No. 2, pp. 135–146.

Souitaris, V. (2002), "Technological trajectories as moderators of firm-level determinants of innovation", *Research Policy*, Vol. 31, No. 6, pp. 877–898.

Spokane, A. R. (1991), "Career intervention", *Education and Training*, Vol. 39, pp. 219–224.

Stasch, S. F. and J. L. Ward (1987), "Some observations and research opportunities regarding marketing of smaller businesses", in G. E. Hills (ed.), Research at the Marketing/Entrepreneurship Interface, Proceedings of the U.I.C. Symposium on Marketing and Entrepreneurship, University of Illinois at Chicago, Illinois, Vol. 15, pp. 39–53.

Stauffer, L. A. and Kirby, A. D. (2003), *The Product Development Needs of Smaller Manufacturing Firms*, Stockholm: International Conference on Engineering Design.

Stoian, M. C., Rialp, A., Rialp, J. and Jarvis R. (2016), "Internationalisation of central and Eastern European small firms: institutions, resources and networks", *Journal of Small Business and Enterprise Development*, Vol. 23, No. 1, pp. 105–121.

Stokes, D. and Fitchew, S. (1997), "Marketing in small firms: towards a conceptual understanding", in Proceedings from the Academy of Marketing 1st Annual Conference, Coral Gables, Florida, pp. 1509–1513.

Stokes, D. and Wilson, N. (2006), *Small Business Management and Entrepreneurship*, 5th ed., London: Thomson, Vol. 27, pp. 78–97.

Stonehouse, G. and Pemberton, J. (2002), "Strategic planning in SMEs-some empirical findings", *Management Decision*, Vol. 40, No. 9, pp. 853–861.

Storey, D. J., (1999), "Human resource management policies and practices in SMEs in the UK: does it really influence their performance?", *Centre for Small and Medium-sized Enterprises, Warwick Business School, University of Warwick*, Vol. 3, pp. 137–165.

Subrahmanya, M. H. B. (2005a), "Small scale industries in India in the globalization era: performance and prospects", *International Journal of Management and Enterprise Development*, Vol. 2, No. 1, pp. 122–139.

Subrahmanya, M. H. B. (2005b), 'Technological innovations in Indian small enterprises: dimensions, intensity and implications', *International Journal of Technology Management*, Vol. 30, Nos. 1–2, pp. 188–204.

Subrahmanya, M. H. B. (2012), "Technological innovation in Indian SMEs: need, status and policy imperatives", *Current Opinion in Creativity, Innovation and Entrepreneurship*, Vol. 1, No. 2, pp. 18–24.

Subrahmanya, M. H. B. (2015), "Innovation and growth of engineering SMEs in Bangalore: why do only some innovate and only some grow faster?", *Journal of Engineering and Technology Management*, Vol. 36, pp. 24–40.

Subrahmanya, M. H. B, Mathirajan, M. and Balachandra K. (2002), "Research and development in small industry in Karnataka", *Economic and Political Weekly*, Vol. 37, No.3, pp. 401–412.

Subrahmanya, M. H. B., Mathirajan, M. and Krishnaswamy, K. N. (2010), "Innovation for SME growth evidence from India", *United Nations University*, Maastricht Economic and social Research and training centre on Innovation and Technology Keizer Karelplein 19, 6211 TC Maastricht, The Netherlands.

Sung, J. and Ashton, D. (2005), *Achieving Best Practice in Your Business, High Performance Work Practices: Linking Strategy and Skills to Performance Outcomes*, London: DTI in association with CIPD, Vol. 67, pp. 265–287.

Swain, A. and Pratihar, S. (2002), "Innovations and challenges in MSME sector", *DRIEMS Business Review*, Vol. 1, No.1, pp. 81–85.

Tabachnick, B. G. and Fidell, L. S. (2001), *Computer-Assisted Research Design and Analysis*, Boston: Allyn and Bacon.

Talukder, M. and Quazi, A. (2010), "Exploring the factors affecting employees: adoption and use of innovation", *Australian Journal of Information Systems*, Vol. 16, No. 2, pp. 1–29.

Tan, C. L. and Nasurdin, A. M. (2010), 'Knowledge management effectiveness and technological innovation: an empirical study in the Malaysian manufacturing industry', *Journal of Mobile Technologies, Knowledge and Society*, available at http://www.ibimapublishing.com/journals/ JMTKS/2010/428053/428053 .pdf (accessed on 12 September 2011).

Taneja, R. (2013), "Challenges of MSME sector in india: an exploratory study", *International Journal of Entrepreneurship & Business Environment Perspectives*, Vol. 2, No. 4, pp. 65–83.

Temin, P. (1979), "Technology, regulation and market structure in the modern pharmaceutical industry", *Bell Journal of Economics*, Vol. 10, No. 2, pp. 429–446.

Thampy A. (2010), "Financing of SME firms in India Interview with Ranjana Kumar, former CMD, Indian Bank; vigilance commissioner, central vigilance commission", *IIMB Management Review*, Vol. 22, pp. 93–101.

Thapa, A., Thulaseedharan, A., Goswami, A. and Joshi, L. P. (2008), "Determinants of street entrepreneurial success", *The Journal of Nepalese Business Studies*, Vol. 5, No. 1, pp. 85–92.

Thornhill, S. (2006), "Knowledge, innovation and firm performance in high- and low technology regimes", *Journal of Business Venturing*, Vol. 21, No. 5, pp. 687–703.

Tidd, J. (2001), "Innovation management in context: environment, organization and performance", *International Journal of Management Reviews*, Vol. 3, No. 3, pp. 169–183.

Tidd, J. and Bessant, J. (2009), *Managing Innovation: Integrating Technological, Market and Organizational Change*, 4th ed., New York: John Wiley & Sons.

Tidd, J. and Hull, F. M. (2006), Managing service innovation: the need for selectivity rather than *"best practice"*, *New Technology, Work and Employment*, Vol. 21, No. 2, pp. 139–161.

Tirkey, M. and Badugu, D. (2012), "Motivation level of employee's in small scale industries in Aligarh district of Uttar Pradesh (India)", *European Journal of Business and Management*, Vol. 4, No.17, pp. 99–114.

Todd, P. R. and Javalgi, R. G. (2007), "Internationalization of SMEs in India", *International Journal of Emerging Markets*, Vol. 2, No. 2, pp. 166–180.

Toftoy, C. N. and Chatterjee, J. (2004), "Mission statements and the small business", *Business Strategy Review*, Vol. 15, pp. 41–44.

Trivedi, J. Y. (2013), "A study on marketing strategies of small and medium sized enterprises", *Research Journal of Management Sciences*, Vol. 2, No. 8, pp. 245–259.

Troilo, M. L. (2014), "Collaboration, product innovation, and sales: an empirical study of Chinese firms", *Journal of Technology Management in China*, Vol. 9, No. 1, pp. 37–55.

Tu, C., Hwang, S. N. and Wong, J. Y. (2014), "How does cooperation affect innovation in micro-enterprises?", *Management Decision*, Vol. 52, No. 8, pp. 1390–1409.

Tunzelmann, N. V. and Acha, V. (2005), "Innovation in low-tech industries", *Business and Management, Innovation, Business Policy and Strategy*, Oxford: The Oxford Handbook of Innovation, pp. 407–432.

Tushman, M. and Nadler, D. (1986), 'Organising for innovation', *California Management Review*, Vol. 28, No. 3, pp. 9–18.

Ughetto, E. (2008), "Does internal finance matter for R&D? New evidence from a panel of Italian firms", *Cambridge Journal of Economics*, Vol. 32, pp. 907–925.

Ullman, J. B. (2001). Structural equation modeling. In: B. G. Tabachnick, & L. S. Fidell (Eds.), *Using Multivariate Statistics*. Boston, MA: Pearson Education.

Utterback, J. M. and Abernathy, W. J. (1975), "A dynamic model of process and product innovation", *OMEGA, The International Journal of Management Science*, Vol., 3, No. 6, pp. 639–656.

Vaessen, P. and Keeble, D. (1995), "Growth-orientated SMEs in unfavourable regional environments", *Regional Studies*, Vol. 29, No. 6, pp. 489–505.

Vatne, E. (1995), "Local resource mobilisation and internationalisation strategies in small and medium sized enterprises", *Environment and Planning*, Vol. 27, pp. 63–80.

Venkatesh, S. and Muthiah, K. (2011), "Power fluctuations -usage of servo voltage stabilizers in industries", *International Journal of Applied Engineering Research, Dindigul*, Vol. 2, No. 1, pp. 283–289.

Verhees, J. H. M. and Meulenberg, T. G. (2004), "Market orientation, innovativeness, product innovation, and performance in small firms", *Journal of Small Business Management*, Vol. 42, No. 2, pp. 134–154.

Vohra, K. (2008), "*Export-Marketing Problems of SMEs: the Case of Ludhiana*", A Dissertation presented in part consideration for the degree of "MA Marketing", The University of Nottingham.

Wang, C., Walker, E. A., & Redmond, J. L. (2007), "Explaining the lack of strategic planning in SMEs: the importance of owner motivation", *International Journal of Organisational Behaviour*, Vol.12, No, 1, pp. 1–16.

Wennekers, A. R. M. and Thurik, A. R. (1999), "Linking entrepreneurship and economic growth", *Small Business Economics*, Vol. 13, No. 1, pp. 27–55.

Wheelen, T. and Hunger, D. (2008), "*Strategic Management & Business Policy*", Boston: Addison-Wesley Publishing Company, Vol. 8, pp. 56–77.

Wickham, P. A. (1997), "Developing a mission for entrepreneurial venture", *Management Decision*, Vol. 35, No. 5, pp. 373–381.

Wiklund J. and Shepherd D. (2003), "Knowledge-based resources, entrepreneurial orientation, and the performance of small and medium-sized businesses", *Strategic Management Journal*, Vol. 24, pp. 1307–1314.

Williamson, I. O. (2000), "Employer legitimacy and recruitment success in small businesses", *Entrepreneurship Theory and Practice*, Vol. 7, pp. 27–42.

Williamson, I. O., Cable, D. M. and Aldrich, H. E. (2002), "Smaller but not necessarily weaker: how small businesses can overcome barriers to recruitment", *Managing People in Entrepreneurial Organizations: Learning from the Merger of Entrepreneurship and Human Resource Management*, Vol. 26, No. 5, pp. 83–106.

Wilson, A. L., Ramamurthy, K. and Nystrom, P.C, (1999), "A multi-attribute measure for innovation adoption: the context of imaging technology", *IEEE Transactions on Engineering Management*, Vol. 46, No. 3, pp. 311–321.

Wilson, L. (1995), "Occupational standards for small firms", *Executive Development* Vol. 8, No. 6, pp. 18–20.

World Bank and the International Finance Corporation (IFC). (2011), "Doing business: India-making a difference for entrepreneurs", http://www.doingbusiness.org/~media/GIAWB/doing%20Business/Documents/Annual-Report/English/ DB11-FullReport.pdf.

Wright, P. M., Gardner, T. M., Moynihan, L. M. and Allen, M. R. (2005), "The relationship between HR practices and firm performance: examining causal order", *Personnel Psychology*, Vol. 58, No. 2, pp. 409–447.

Xie, X. M., Zeng, S. X. and Tam, C. M. (2010), 'Overcoming barriers to innovation in SMEs in China: a perspective based cooperation network', *Innovation, Management, Policy & Practice*, Vol. 12, No. 3, pp. 298–310.

Yin, X. and Zuscovitch, E. (1998), "Is firm size conducive to R&D choice? A strategic analysis of product and process innovations", *Journal of Economic Behavior and Organization*, Vol. 35, No. 2, pp. 243–262.

Yun, J. J. (2017), *Business Model Design Compass: Open Innovation Funnel to Schumpeterian New Combination Business Model Developing Circle.* Cham, Switzerland: Springer (Chapter 2).

Zahra, S., Ireland, R. and Hitt, M. (2000), 'International expansion by new venture firms: international diversity, mode of market entry, technology learning, and performance', *Academic of Management Journal*, Vol. 43, No. 5, pp. 925–950.

Zairi, M. (1994), 'Innovation or innovativeness, results of a benchmarking study', *TQM Magazine*, Vol. 5, No. 3, pp. 10–16.

Zaugg, R. and Thom, N. (2003), "Excellence through implicit competencies: human resource management-organizational development-knowledge creation", *Journal of Change Management*, Vol. 3 No. 3, pp. 199–211.

Zeng, S. X., Xie, X. M. and Tam, C. M. (2010), "Relationship between cooperation networks and innovation performance of SMEs", *Technovation*, Vol. 30, No. 3, pp. 181–194.

Zohar, D. (1980), "Safety climate in industrial organizations: theoretical and applied implications", *Journal of Applied Psychology*, Vol. 65, No. 1, pp. 96–102.

Appendix I: Technology innovation questionnaire

Organization's Name & Address				
Respondent's Name & Designation				
Respondent's E-Mail Address				
Respondent's Contact No. / Fax No.				
Type of Organization	Manufacture			
	Service			
	Commerce and Trade			
	Any Other (Please Specify)			
ANNUAL TURNOVER				
YEAR OF INCEPTION				
NUMBER OF EMPLOYEES	a) 1-99	b) 100-199	c) 200-499	d) >500

(Signature of Respondent)
WithSeal

SECTION I

PROBLEMS FACED BY MSMEs				
1.Which of the following factors related to Human Resource Management (HRM) are responsible for poor performance of MSMEs?				
1	Improper recruitment and selection practices	Not at all	To some extent	Reasonably well	To a great extent
2	Difficulty in finding suitable manpower in the labor market	Not at all	To some extent	Reasonably well	To a great extent
3	Lack of job specific training to update knowledge about latest machinery	Not at all	To some extent	Reasonably well	To a great extent
4	Lack of formal orientation programs for new employees	Not at all	To some extent	Reasonably well	To a great extent
5	Lack of interest of staff to attend training programs	Not at all	To some extent	Reasonably well	To a great extent
2. How Strategic Financial Management is affected by following factors in an organization?				
1	Difficulty in getting loans	Not at all	To some extent	Reasonably well	To a great extent
2	Lack of systematic budget system (Earmarking of funds in advance for specific activities)	Not at all	To some extent	Reasonably well	To a great extent
3	Insufficient credit for meeting requirements of routine operations	Not at all	To some extent	Reasonably well	To a great extent
4	Lack of use of specific financial strategies for effective utilization of funds	Not at all	To some extent	Reasonably well	To a great extent
3.Which of the following aspects are responsible for a lesser amount of Technology Dynamism in MSMEs?				
1	Lack of appropriate raw materials at reasonable prices	Not at all	To some extent	Reasonably well	To a great extent
2	Lack of budget allocation for R&D initiatives and new product development projects	Not at all	To some extent	Reasonably well	To a great extent
3	Lack of state of the art process technology in use	Not at all	To some extent	Reasonably well	To a great extent
4	Lack of increase in product variants/new products as a result of in-house technology up-gradation	Not at all	To some extent	Reasonably well	To a great extent

(Continued)

PROBLEMS FACED BY MSMEs				
5	Lack of improvement in existing product features due to technology up-gradation	Not at all	To some extent	Reasonably well	To a great extent
4. To what extent the given issues contribute towards poor Systematic Planning?					
1	Lack of analyses of own-potential before starting new projects	Not at all	To some extent	Reasonably well	To a great extent
2	Lack of use of planning tools, especially operation management tools for better production planning and control	Not at all	To some extent	Reasonably well	To a great extent
3	Lack of knowledge of entrepreneur regarding various government schemes for MSMEs	Not at all	To some extent	Reasonably well	To a great extent
5. How does the listed Physical Infrastructure factors diminish the performance of MSMEs?					
1	Lack of cheap and reliable power supply	Not at all	To some extent	Reasonably well	To a great extent
2	Lack of good rail and road infrastructure	Not at all	To some extent	Reasonably well	To a great extent
6. How the following factors related to Market Research are associated with the poor performance of MSMEs?					
1	Lack of use of market function as an important activity to collect information on product, customer etc.	Not at all	To some extent	Reasonably well	To a great extent
2	Lack of proper marketing to promote products	Not at all	To some extent	Reasonably well	To a great extent
3	Lack of up-to-date marketing knowledge of different marketing segments	Not at all	To some extent	Reasonably well	To a great extent
7. How does the following Organizational Culture issues affect the performance of MSMEs?					
1	Poor understanding of organization's mission and goals	Not at all	To some extent	Reasonably well	To a great extent
2	Inappropriate working environment in the organization	Not at all	To some extent	Reasonably well	To a great extent
3	Inappropriate organizational practices and policies	Not at all	To some extent	Reasonably well	To a great extent

(Continued)

PROBLEMS FACED BY MSMEs				
4	Lack of career development opportunities for skilled employees	Not at all	To some extent	Reasonably well	To a great extent

8. To what extent the lack of various Cooperation Networks effect the performance of MSMEs?

1	Lack of cooperation between MSMEs and government agencies	Not at all	To some extent	Reasonably well	To a great extent
2	Lack of inter-firm cooperation for MSMEs	Not at all	To some extent	Reasonably well	To a great extent
3	Lack of cooperation between MSMEs and intermediary institutions	Not at all	To some extent	Reasonably well	To a great extent
4	Lack of cooperation between MSMEs and research organizations	Not at all	To some extent	Reasonably well	To a great extent

9. How does the Poor Geographical Location of MSMEs reduce their performance?

1	Poor connectivity of our state with other parts of the country/outside region	Not at all	To some extent	Reasonably well	To a great extent
2	Absence of large scale industrial sector in the region	Not at all	To some extent	Reasonably well	To a great extent
3	Lack of access to global markets	Not at all	To some extent	Reasonably well	To a great extent
4	Difficulty for regular customers to get to our location on a regular basis	Not at all	To some extent	Reasonably well	To a great extent
5	Too many number of similar business located nearby	Not at all	To some extent	Reasonably well	To a great extent

SECTION 2

INPUT RESEARCH PARAMETERS				

1. (a) Which type of Entrepreneurial Capabilities are desired for taking TI initiatives to improve manufacturing performance of MSMEs?

1	Good education level of entrepreneur (both general and business specific)	Not at all	To some extent	Reasonably well	To a great extent
2	Ability to make effective decisions pertaining to business activities	Not at all	To some extent	Reasonably well	To a great extent
3	Knowledge regarding various government schemes for MSMEs	Not at all	To some extent	Reasonably well	To a great extent

(Continued)

	INPUT RESEARCH PARAMETERS				
4	Technical competencies of entrepreneur like competency in operating all machines, quality control tools etc.	Not at all	To some extent	Reasonably well	To a great extent
5	Entrepreneur Training	Not at all	To some extent	Reasonably well	To a great extent
6	Strategic decision making in identifying right kind of business and market	Not at all	To some extent	Reasonably well	To a great extent
7	Tactic knowledge obtained through prior working experience	Not at all	To some extent	Reasonably well	To a great extent
1. (b) What kind of awareness shall be helpful in implementing TI initiatives for improved manufacturing performance?					
1	Alertness and formal business planning of different management areas like finance, H.R, logistics etc.	Not at all	To some extent	Reasonably well	To a great extent
2	Knowledge about various financial schemes and procedures to be followed to obtain loans etc.	Not at all	To some extent	Reasonably well	To a great extent
3	Ability to seize opportunities from market and using appropriate strategies in commercializing new product	Not at all	To some extent	Reasonably well	To a great extent
4	Awareness about new production technologies, machines, equipments etc.	Not at all	To some extent	Reasonably well	To a great extent
5	Strong emphasis on the development of new innovative products or improved products	Not at all	To some extent	Reasonably well	To a great extent
6	Strong emphasis on R&D, technological leadership and innovations	Not at all	To some extent	Reasonably well	To a great extent
7	Ability to introduce new products services, techniques and technologies	Not at all	To some extent	Reasonably well	To a great extent
2. (a) Which kind of Technology Infrastructure is required for improving manufacturing performance through TI initiatives?					
1	Appropriate raw materials at reasonable prices	Not at all	To some extent	Reasonably well	To a great extent

INPUT RESEARCH PARAMETERS				
2	Appropriate process technology infrastructure i.e. state of the art production machinery, equipments, tools etc. to meet the changing market demands	Not at all	To some extent	Reasonably well	To a great extent
3	Technical knowledge and infrastructure to do business operations with Information Systems (e-purchasing, use of RFID and bar codes etc.)	Not at all	To some extent	Reasonably well	To a great extent
4	Internet for marketing and promoting products	Not at all	To some extent	Reasonably well	To a great extent
5	Implementation of new types of production processes	Not at all	To some extent	Reasonably well	To a great extent
6	Acquiring manufacturing technology and skills entirely new to the firm	Not at all	To some extent	Reasonably well	To a great extent
7	Organization's ability to transform the results of R&D into products that meet market needs	Not at all	To some extent	Reasonably well	To a great extent
8	Softwares for designing and production related tasks for product development	Not at all	To some extent	Reasonably well	To a great extent
2. (b) What type of financial support is required for improving manufacturing performance through TI initiatives?				
1	Financial budget, especially for R&D initiatives and new product development projects	Not at all	To some extent	Reasonably well	To a great extent
2	Systematic budget system i.e. earmarking of funds in advance for specific R&D initiatives and projects	Not at all	To some extent	Reasonably well	To a great extent
3	Specific financial strategies for effective utilization of available funds	Not at all	To some extent	Reasonably well	To a great extent
4	Sufficient credit for meeting requirements of routine operations	Not at all	To some extent	Reasonably well	To a great extent
5	Getting loans from banks for technology up-gradation initiatives	Not at all	To some extent	Reasonably well	To a great extent

(*Continued*)

INPUT RESEARCH PARAMETERS				
3. Please indicate the level of Organizational Culture and Climate which is essential for improving manufacturing performance of MSMEs?				
1	Skilled manpower to increase competitiveness and sustainable growth	Not at all	To some extent	Reasonably well	To a great extent
2	Intensity of R&D personnel (ratio between the number of full-time employees engaged in R&D and total number of employees in the organization)	Not at all	To some extent	Reasonably well	To a great extent
3	Extent of training given to employees to transmit knowledge and skills of requisite quality	Not at all	To some extent	Reasonably well	To a great extent
4	Formal reward structure to motivate employees (financial rewards, promotions and carrier development opportunities)	Not at all	To some extent	Reasonably well	To a great extent
5	Employee empowerment to take responsibility for technological innovation initiatives	Not at all	To some extent	Reasonably well	To a great extent
6	Management support for technological innovation initiatives	Not at all	To some extent	Reasonably well	To a great extent
7	Education level/qualification of production personnel	Not at all	To some extent	Reasonably well	To a great extent
8	Extent of using market and customer feedback into the innovation processes	Not at all	To some extent	Reasonably well	To a great extent
4. Which of the following Government Initiatives are helpful for improving manufacturing performance of MSMEs?				
1	Providing support in acquiring latest technology, quality certification and marketing assistance	Not at all	To some extent	Reasonably well	To a great extent
2	Providing training to employee at government institutes like tool rooms	Not at all	To some extent	Reasonably well	To a great extent
3	Providing lab facilities for R&D initiatives at subsidized rates	Not at all	To some extent	Reasonably well	To a great extent
4	Free or subsidized information regarding latest trends and technologies in relation to government regulations	Not at all	To some extent	Reasonably well	To a great extent

(Continued)

INPUT RESEARCH PARAMETERS				
5	Increase in important policies and measures to support innovation initiatives in MSMEs	Not at all	To some extent	Reasonably well	To a great extent
6	Allocating funds for R&D initiatives in MSMEs	Not at all	To some extent	Reasonably well	To a great extent
7	Tax policies for MSMES to encourage them to speed up technological and new product development projects	Not at all	To some extent	Reasonably well	To a great extent
8	Providing cheap and reliable power supply to the MSMEs	Not at all	To some extent	Reasonably well	To a great extent
9	Providing good rail and road infrastructure in the region	Not at all	To some extent	Reasonably well	To a great extent

MANUFACTURING PERFORMANCE PARAMETERS

1. Please indicate the level of improvement in Product Performance due to TI initiatives in your organization?

1	Increase in cost/benefit ratio of company's products because of technology up-gradation initiatives as compared to competitor's products	< 5%	5-10%	10-30%	> 30%
2	Improvement in product life cycle of the products	Not at all	To some extent	Reasonably well	To a great extent
3	Reduction in the time (from inception of idea to launch) to develop new products	Not at all	To some extent	Reasonably well	To a great extent
4	Reduction in the cost of production	Not at all	To some extent	Reasonably well	To a great extent

2. Up to what extent Innovation Performance is enhanced due to TI initiatives in your organization?

1	Increase in proportion of new products as a percentage of total products over the last 3-5 years	< 5%	5-10%	10-30%	> 30%
2	Improvement in technical characteristics and features of existing product range	Not at all	To some extent	Reasonably well	To a great extent
3	Implementation of new processes as a result of technology up-gradation initiatives	Not at all	To some extent	Reasonably well	To a great extent
4	Extension in range of products in the last 3-5 years	Not at all	To some extent	Reasonably well	To a great extent
5	Adaptation of the basic and key technologies	Not at all	To some extent	Reasonably well	To a great extent

(Continued)

INPUT RESEARCH PARAMETERS				
3. Please indicate in terms of following parameters the improvement in sales due to TI initiatives in your organization?				
1	Mean sales profitability (increase in profit margins) over the last 3-5 years	< 5%	5-10%	10-30%	> 30%
2	Sales improvement due to new products as a percentage of total sales	< 5%	5-10%	10-30%	> 30%
3	Increase in market share because of new products	< 5%	5-10%	10-30%	> 30%
4	Penetration into new markets	Not at all	To some extent	Reasonably well	To a great extent

Appendix II: Letters of support from various manufacturing organizations

Following are the reputed manufacturing organizations offered to extend their whole-hearted support for the research work. These organizations had shown their willingness to participate in the research work and the letters of supports are listed below:

GLOBAL ENGINEERS & INNOVATORS

Works : E-24, Industrial Area, Phase VII, SAS Nagar, Mohali, Punjab (INDIA)
Ph. : 0172-2210621, 98148-33621, e-mail : avtarglobal@yahoo.co.in

TO WHOM IT MAY CONCERN

We are pleased to know that Mr. Davinder Singh is pursuing his Ph. D. research work on problems related to small firms and evaluation of Technology Innovation Initiatives of MSMEs. His title of research work *"An Evaluation of Technological Innovation Initiatives of MSMEs in North India"* is highly related with the work undertaken by us in the last few years, since we are also committed towards technology development.

Model to encourage technological innovation in small firms is long overdue since small manufacturing units are facing enormous problems related to skilled manpower, technology infrastructure, finance, location, power supply, raw material etc.

We wish Mr. Davinder Singh all the best for his research on problems related to small firms and evaluation of Technology Innovation Initiatives of MSMEs and assure to provide him all support for the completion of the research work.

Signature:

Name: AVTAR SINGH

Designation: Prop.

TIN NO. 03282101404   Mob.: 98721-82710, 98786-07700
                                                        E-Mail.: bhstools@yahoo.com

BHS TOOLS

CUTTING TOOLS MANUFACTURERS:- All Types of Broaches

SANOURI MANDI ROAD, HIRA BAGH, PATIALA

Ref. No............   **TO WHOM IT MAY CONCERN**   *Dated*.........................

Mr. Davinder Singh has performed a case study related to
PROBLEMS   AND   TECHNOLOGY   INNOVATION
INITIATIVES OF MSMEs in our industry. We wish him a
great success for the completion of his research work.

Signature:

Name:

Designation:

Index

For Product Safety Concerns and Information please contact our EU
representative GPSR@taylorandfrancis.com
Taylor & Francis Verlag GmbH, Kaufingerstraße 24, 80331 München, Germany